Jackie Moffat is a Londoner born and bred, but in 1982 she moved with her husband Malcolm to the Eden Valley in Cumbria, where they still live and farm. As well as breeding Manx Loghtans – a rare breed of sheep originating from the Isle of Man – Jackie writes a column for Cumbria & Lake District Life magazine. She is the author of the bestselling *The Funny Farm*. To learn more about life at Rowfoot, visit www.jackiemoffat.co.uk

Also by Jackie Moffat

The Funny Farm

and published by Bantam Books

Sheepwrecked

Jackie Moffat

BANTAM BOOKS

LONDON · TORONTO · SYDNEY · AUCKLAND · JOHANNESBURG

TRANSWORLD PUBLISHERS
61–63 Uxbridge Road, London W5 5SA
a division of The Random House Group Ltd
www.booksattransworld.co.uk

SHEEPWRECKED
A BANTAM BOOK: 9780553817768

First published in Great Britain
in 2006 by Bantam Press
a division of Transworld Publishers
Bantam edition published 2007

Mixed Sources
Product group from well-managed
forests and other controlled sources
www.fsc.org Cert no. TT-COC-2139
© 1996 Forest Stewardship Council
FSC

For Eleanor and Phoebe,
two very special girls, with love

Contents

Acknowledgements

Thanks, as always, to my husband Malcolm for his continued unstinting support and encouragement. The publication of this book proves that I have been writing properly from time to time and not just playing on t'internet.

Special thanks to Patrick for his foreword and to Jane, my old partner in crime from schooldays, for her brilliant photos. Other fantastic pictures have been kindly provided by Lionel, who has become a friend as a direct consequence of his reading *The Funny Farm*. Lionel, thank you.

Grateful thanks also to Simon, my editor at Bantam, and to all the Transworld team who are wonderfully tolerant of my ignorance about matters they all understand so well. Thank you all.

I should also like to express my appreciation to all the people who have made contact with me since *The Funny Farm* was published – *Sheepwrecked* is for you all and I hope you enjoy it just as much.

Foreword

Early in life the author's mother must have recognised her daughter's revolutionary tendencies and that the saying 'You should never mess with a red-head' is unfortunately true. Son-in-law having broken his back in a fall from a horse on holiday at Morecambe Bay, one detects her despair at the announcement that said daughter, plus damaged husband, planned to emigrate to a disused smallholding in ultimate Cumbria to pursue the Good Life in preference to their more conventional habitat of Walton-on-Thames.

The ensuing story is told, as she speaks, with humour and self-deprecation. It touches on the wider lunacies of central authority and the idiosyncrasies of newly discovered agricultural neighbours.

Knowing Jackie, who, with her husband Malcolm, has been my occasional friend for twenty years, has greatly enriched my life. Her scepticism of official doctrine is healthy, unusually free from spite, and completely in tune with the countryside to which she has become accustomed at a time when its best traditions of the care of livestock and the soil have been bleeding to death.

This book will encourage those who believe that if

you have sufficient to live on, a roof over your head and conviction about what you are doing, plain financial wealth in the purlieus of Walton-on-Thames is irrelevant.

Jackie is a brave and beautiful woman married to a man who has supported her through thick and thin, having learnt from his days in the police force that it is wise to 'come along quietly'.

The girl done good.

Patrick Gordon-Duff-Pennington
Muncaster Castle, Ravenglass, Cumbria

Prologue

*The contents of a woman's handbag are a pretty
accurate barometer of her lifestyle. A quarter of a
century (gosh, how time flies when you're enjoying
yourself) ago, mine contained a train season ticket –
outrageously priced; keys – so handy for disfiguring
muggers; various lipsticks, spare tights and hair
spray. Now, a quick rummage reveals a box of a
hundred Elastrator rings for castrating lambs
painlessly (orange, so that you can see them in the
long grass, and as far as the 'painless' goes, well,
you'd have to ask a tup lamb really – expect a high
octave reply); a very un-cool mobile phone the size
of a shoe box, and a leaflet about domestic wind
turbines. Would you trust a woman who carries
such things about with her? No, me neither.*

S EASONS SHIFT. DAILY tasks and patterns change and
living in harmony with the land means that you, as
much as the wildlife or the hedgerows or the skeins of
geese overhead, are part of that unchanging seasonal
rhythm. Nature calls the tune and the countryman keeps

time with her seasonal music. There's no option: try telling snowdrops to come up in summer, or ewes to mate in spring. The man on the top deck of the Clapham omnibus, as he cruises the townscape, is not part of those changes in the same way as his country peers and that is probably the biggest, most important difference between their parallel lives. It may also be the benefit of staying put in Clapham.

As far as seasonal changes go, mere mortals like me can do little about it except watch, groan and grumble about the weather. Here in Cumbria there is a joke – they have an odd sense of humour, Cumbrians – that we get nine months of winter and three of bad weather. I don't think that is especially true but I do confess to a wry smile – some might say it's not a smile at all but a pretty unpleasant smirk – when it's sunny here and pelting in Basingstoke.

As far as the changing patterns of life go, ours have changed as often as the emphasis of life at Rowfoot has shifted.

In the beginning, there was the Word and ours was Construction. Or deconstruction, as we concentrated on some sort or other of earth-moving enterprise, mostly inside the house. At the height of activities, we came to regard proper wardrobes as a petty bourgeois affectation – after all, we had managed for the best part of a year with Pickfords cardboard ones – and plaster dust had become an essential ingredient in our tea and coffee.

As Malcolm toddled off each morning for another day at Newton Rigg learning how to castrate calves, milk pigs, shear goats and tinker with tractors, those being a few of the unlikely topics he claimed to be studying while the younger students concentrated on getting their PhDs in beer drinking, I tooled up for another day as brickie's mate. Ted was chief brickie and a long-standing friend. Indeed it was Ted who first suggested to us, on one of our house-hunting forays, that we should have a look at Rowfoot. Sometimes I think Ted has an awful lot to answer for. He dished out orders and I executed them. Under his direction, I pulled plaster off the walls, found the odd rats' nest lurking in the stonework where normal people have cavity wall insulation, and dug out floors. I can still see my mother standing in the doorway of the old dairy, wondering precisely why her only offspring needed great sandstone slab shelves when she was perfectly happy with blue Formica ones in Barnes. 'I just don't understand you, Jacqueline, moving from a perfectly dry house in Walton-on-Thames to a damp one in Cumbria,' she said in bewildered tones. At times like this, she became more convinced than usual that she had been sent home from hospital with the wrong baby.

Somehow, there seemed little point in even beginning to explain about the views, the space, the fields, the livestock, the imminent translation of the self-sufficiency dream into lip-smacking reality. Besides, I knew that it wasn't really the damp that bothered mum, it went far deeper than that: the real problem with this country living lark was the lack of a Marks and Spencer within walking distance.

The day before, I had cursed at a sudden lack of cheese; the wall-rats' descendants might conceivably have been responsible but allusions to intramural rodents might have upscuttled a lady who had been a spectator of life's unfolding tapestry from the vantage point of a blue kitchen chair (to match the Formica) as mice scurried about on the floor of a rambling Edwardian house in deepest suburbia. As an infant, I found it amusing to flick small morsels of dry bread at them as they scurried about my playpen, an activity whose entertainment value my mother never fully appreciated.

'I'll have to nip to Armathwaite for some cheese,' I said.

Mother, pouncing on a chance to escape her un-civilised surroundings, thick with plaster dust and builders' rubbish, exclaimed excitedly, 'Armathwaite? Does it have a Marks and Spencer's?'

Briefly, Armathwaite beckoned like a Lorelei, glittering and irresistible. It was a shame to disillusion her but it had to be done. No Marks and Sparks: Armathwaite's allure faded immediately.

Back then, the days slid easily into one another: on the five weekdays Ted and I toiled away and at the weekends Malcolm, Sara (Malcolm's youngest daughter), Ted and I all got stuck in to the major jobs together. Sometimes we roped in visitors, family members in transit and innocent passers-by who had set out to do the East Cumbria Countryside Project walk from Armathwaite to Ainstable but were hijacked en route. Usually we released them after a couple of days' hard labour in captivity. It was a strangely satisfying time.

Anxious to restore what we could, we replaced the modern tiled fireplace in the sitting room with a magnificent stone one, hewn from the recycled dairy floor. Its construction had been rather more straight-forward in the imagining than in the execution. 'We'll use the front gatepost for a lintel,' suggested Ted airily, 'it's good dressed stone, that. Just replace it with the one down by the hedge in the bottom field.' As the fields had been enlarged, boundaries had been taken out, and the lone rough and redundant pillar referred to stood an idle monument to a past of smaller fields and correspondingly smaller agricultural vehicles. We could widen the gateway, too, Ted reasoned, and that would be of inestimable benefit: all sorts of wide and heavy machinery could get into the yard then. Why? What could we possibly need with wide and heavy machinery? 'You never know,' replied Ted darkly and carried on digging.

There was just the small matter of how to shift our twin peaks. Tom, our dairy farming neighbour, at the helm of a JCB and a front-loader (I had thought this term pertained only to washing machines, but apparently not) saw to the one in the field, but the journey of the monster gatepost that guarded the entrance to Rowfoot to its new home in the sitting room presented a different challenge altogether. Ted eventually rigged up a con-traption to lever it into place, incorporating a small flatbed trolley and a winch of sorts; it defied accurate description but it was just the sort of thing that made Heath Robinson great. ACRO props held the lintel in place and Sara, whose bedroom was immediately above the half-completed structure, was told very firmly not to jump about upstairs.

Upstairs, that other country nearer heaven and above damp-course injections and smashed plaster, offered another heady possibility: an en-suite bathroom. This was Malcolm's idea and, frankly, I could not see it ever going beyond the crumpled graph paper stage.

Our predecessors at Rowfoot had put in a proper bathroom – a Good Idea. Another bedroom had been added too – also a Good Idea. They had been left with an odd L-shaped space and had stuck a door on it: not a good idea at all as it was exactly the kind of space that invites junk to parties, then encourages it to sleep on the floor and never leave. I had visions of various sorts of detritus claiming squatters' rights there in the years to come but it seemed an unlikely space for a fancy bathroom, especially if you wanted a pukka bath as opposed to one of those French bum-baths (as long as your bum is no larger than a standard size 14) in it. But the graph paper, insisted Malcolm, could not lie. It would go in. At least, it would go in if we halved the size of the airing cupboard, took off the existing door, plasterboarded over the hole where the door had been, knocked a new hole through the old outside wall of the house to create a new opening from our bedroom and stuck a door on that instead. Ah yes, of course, I said, why didn't I think of that, in my best encouraging-wifely he'll-forget-all-about-it-in-a-day-or-so voice.

He started on the hole the next day.

To make a very narrow doorway indeed, an irregular space roughly two metres high by five metres wide had to be made in the old outside wall that was itself fully a metre thick. Boulders the size of Dufton Pike were carried, rolled and manhandled down the stairs and still the house did not fall down. This was heartening.

We relayed some of this information to our erstwhile bank manager in Putney, certain that he would see the joke. He, after all, had conceded that he didn't come across too many surveys of seventeenth-century farm-houses in the course of Putney's daily grind. He said he thought we were pioneer spirits. He may have meant that we were bonkers but was much too polite to say so.

Ted thought that four concrete lintels might just suffice over the new doorway and his next thought was that it would be quicker (than what?) if I went to Carlisle to fetch them in the Land Rover.

I had made many journeys to Coulthards of Carlisle, suppliers of concrete lintels, other toys and mysteriously named 'Sundries' to the building trade. Whilst I never fully established what exactly constituted a Sundry, I did discover during these forays that Coulthards had some very helpful staff and one or two complete wastrels with delusions of adequacy, the kinds of blokes who were depriving remote Cumbrian settlements of their resident idiots.

The most memorable of the latter sucked heavily on a fag hanging perpetually and perilously from one corner of his mouth and talked out of the other corner with appropriate economy of movement in the lip department. He did not like waste, particularly of his own energy, and he tried his level best not to expend any of this irreplaceable commodity unless it was absolutely necessary. Years of research had resulted in his development of a sophisticated modus operandi: lots of sucking on the fag and very little of anything else. To his immense satisfaction he had discovered that his inertia slowed things down to such an intolerable degree that most customers felt impelled to speed up the process by lending a hand and lifting their own lintels, bags of concrete and, of course, Sundries.

Such low tactics invited a cunning counter-attack, delivered with feminine guile and involving no fag-sucking at all. And here it is: I have not patented it and you may use it freely and at your own discretion, just so long as the victim is deservingly weaselly. The first rule is: do not, under any circumstances, wear jeans, steel-capped boots or any other clothes in which you might be expected to lift lintels, bags of concrete or – perish the thought – Sundries. Wear a short black skirt instead. And very high heels. As you drive into the storage yard, wave your pink duplicate order form and instruct (it may help if you think of your old hockey mistress while you do this): 'Four concrete lintels, two bags of cement, in the back please.' And then, very grudgingly, you should slide the window across – our old Land Rover had those sticky, supposed-to-be-slidy windows that predated even the roll-down sort, never mind the electronic ones, and add, in the most imperious tone you can muster, 'Would you require the tailgate lowered?'

It is up to you how you proceed now. You might deal with tailgates yourself or you could just issue instructions on tailgate lowering with special reference to Land Rovers . . . it worked for me.

*   *   *

The extra bathroom was taking shape nicely, even if we
had truncated the airing cupboard in the main one. On
reflection, this was no bad thing, as heating an airing
cupboard that vast must surely have been a wasteful
use of energy, but the modifications did mean that
we had to resite the immersion switch. And this had to
be approved by the building regs inspector, a very
important district council person with whom you
disagreed at your peril. There was one inspector in
particular who struck fear and awe into the hearts of
builders and householders alike. I'll call him Mr Price,
as this name caught my attention in a newspaper
report recently. The real Mr Price of New South Wales,
Australia, was butchered by his wife Katherine in 2003;
the charming Kate, being deft with a cleaver, stabbed
her erstwhile love a total of thirty-seven times before
decapitating him, skinning him and eventually cooking
'parts of his flesh' although, regrettably, we are not told
exactly which parts. I think we can guess, though, don't
you? She then fed him to her children, with a certain
culinary flourish and the obligatory vegetables and
gravy. Katherine is now languishing in an Australian
jail and her children are, I am sure, receiving a more
balanced diet – or at least one prepared by a more
balanced cook.

But that is to digress; back to my putative Mr
Price whose diligence and assiduity are such that his
continued immunity from machete-wielding females
cannot, I fear, be guaranteed.

He arrived and inspected the floors we had taken out

and those we had put in. He looked at the walls, demolished and molished alike. He approved of both plumbing and plasterwork: we were on a roll. The new bathroom he pronounced to be in full compliance with all his knavish regulations and 'very nice' (he was not a man much given to extravagant expressions of appreciation). 'I'll just have a look at the airing cupboard on the other side,' he declared, in tones suggestive of dark arts and weird satanic practices. He opened the cupboard door with due solemnity and peered inside.

'Hmm.'

I quivered a little.

He fixed me with a gimlet eye. Well, two gimlet eyes really; he wasn't a Cyclops or anything like that.

And then very slowly, very carefully and of course fully dressed, he stepped into the bath and stretched his left arm as far as he could, in an effort to reach the immersion switch in the airing cupboard.

He looked quite spectacularly silly.

To his dismay, he could not quite touch the switch. If he had been able to, he explained, the house would quite possibly have to be knocked down altogether as anyone bathing recklessly might have found that, with minimal effort and very long arms, they could electrocute themselves. Since his own search for imperfection had been in vain, he quit the bath and stalked down the stairs, a thwarted man.

The electrician responsible for the switch-resiting project looked very smug indeed and toiled away elsewhere in the house. A wraith-like youth with horn-rimmed specs and an odd taste in pipes, he whistled tunes from *Oklahoma* and grunted in the way that Tamworth pigs, teenaged boys and tennis players have perfected, but he never said very much. He reserved his rare bursts of conversation to abuse the workmanship of another, earlier electrician, who was alternately a – well, I can't really repeat what he called him. He meant that the chap was a charlatan and a rogue but those were not, as I recall, the precise terms he used to convey his opinion.

Still, we were fortunate to have electricity in Ainstable at all. Had things gone to the Great Master Plan of one Ted Proud, whose parents had resided at Rowfoot at the turn of the century, we would not have had electricity at all. Ted was all against the leccy. Why? It might just have been that he was the most dreadful Luddite. On the other hand – and I incline towards this explanation myself – it might have had something to do with the fact that Ted had the paraffin franchise for the area.

Ted's son Eddie, though, was responsible for reminding us that Sundays were special. Every Sunday morning for several weeks after we moved in, a cattle-feed bag materialised on the back door step; it was filled not with barley or 'cake' but with a selection of seasonal vegetables. Eventually, the Turnip Fairy was identified as Eddie, who explained his kindness with a gruff 'Well, I thought you hadn't had chance to get your garden goin' yet.'

After that we persuaded him to share a glass of elderberry wine with us when he brought the Sunday veg, especially as he had a family connection with

Rowfoot. The elderberry was an amusing little hooch, made from the fruits of bushes in the park at the back of our Surrey house and matured in the loft there. One or two bottles didn't make it northwards: 'Daddy, daddy, there's blood coming down the walls . . .' cried Sara in terror one morning, a bottle or three of lively elderberry having exploded in the loft and made their passage downwards with devastating effect. Those which survived intact we transported to Cumbria.

One Sunday, leaning against the half-plastered wall – the wall being semi-denuded of its coating in preparation for the damp-course injection rather than having indulged too freely in the elderberry wine – and gesturing to the opposite side of the kitchen, Eddie said, 'There used to be a bath in yon corner.' The next day, we were smashing yet more plaster and what fell on to the floor amidst the dust and a brace of rat ribs? A bath plug. No use, though – imperial width, you see, not metric.

On other Sundays, Eddie talked of childhood visits to Rowfoot, remembering the pear tree as bearing 'fruit like little bullets, hard and sour. Then, suddenly, they ripened and then they were soft and juicy and sweet . . .' No change there: the pear tree's tiny fruits were still as volatile and as tasty as they had been fifty years ago. And they made some pretty good wine, too.

Eddie's tractor used to describe a pleasingly erratic path up the road after the elderberry wine had done its stuff. The tractor was as a calendar on other days of the week too: on hunting days – Mondays, Wednesdays, Fridays and Saturdays – it achieved third gear and a giddy 15 mph. On the intervening days its progress

12

was of a more leisurely nature. If, though, the foxes became particularly troublesome and the Bewcastle went a-hunting on extra days, the tractor's progress would accelerate accordingly. So you had to be very careful if you relied on the velocity of Eddie's Massey Ferguson to identify which day of the week it was.

As the seasons shifted, our patterns changed too.

At the end of that summer's grazing licence we reinstated fencing, enabling us to divide the land back into manageable fields rather than having it all lumped within one ring fence.

Then came Norman with his Crane of Many Colours and his Mole Plough to lay water pipes. Norman lives at the top of the village in a clearing crammed with mechanical experiments and surrounded by a tangle of trees. He is quite simply a genius with all agricultural machinery and much else besides. Norman's only real failing is his sense of colour co-ordination. He and Laurence Llewelyn-Bowen would never get on as Norman regards paint not as a fashion consideration but as a utilitarian substance that protects his creations from the elements. Or at least that seems to be the most satisfactory explanation for Norman's Crane of Many Colours, its component parts splashed with an eclectic mix of vibrant hues: Kermit the Frog Green, Father Christmas Red, Microsoft Toolbar Blue and Banana Custard Yellow are all there.

The tractor-mounted Mole Plough is another of Norman's creations. The term 'mole plough' is self-explanatory if you think about it long enough. A brightly painted (natch) tractor tows a ferocious blade, slicing into the earth; at the deepest point is the 'mole', burrowing the course for the water pipes. It saves digging massive trenches with spades and laboriously filling them in, causes little disruption to the growing surface of the land and is devilishly accurate. A very neat implement indeed, is a mole plough, though an organic farmer would probably prefer to use a real, smooth-jacketed mole. But training him to follow the maps between the troughs might prove a bit tricksy.

There didn't seem to be a good reason for the Crane of Many Colours to come with the Mole Plough; perhaps they were just good friends. But as summer turned to autumn and then to winter water pipes were sunk, troughs were installed, ballcocks were fitted and sheep were bought.

Shepherding became another daily ritual, and for the most part one day was much like another. Nor did the arrival of the dairy cows change anything to start with. They needed milking every morning and every evening of every day. Only when we began to make butter and cheese did the week start to take on a different shape.

We processed everything on site and sold no milk in unseparated form. So, after every milking, the milk, still at blood heat, passed through a separator. At first we had a hand-wound model, ensuring that I developed impressive biceps on my right arm. Malcolm said that ours was absolutely no different from the one his Auntie Polly had had fifty years earlier so it is probably fair to assume that Auntie Polly had impressive biceps too.

And like Auntie Polly we made butter. Surplus cream was heat-treated each night in the microwave and stored in the dairy fridge for making into cream cheese on Wednesdays or churning into butter on Fridays.

The cream cheese production was a straightforward process. I cannot explain the logic of it, though Malcolm says he can – it is because, apparently, the system was devised by a woman. Me. The cream was heated to 70 degrees centigrade and then cooled to 98 degrees Fahrenheit. It seemed obvious enough to me: the touch pad on the widescreen microwave – it was state-of-the-art at the time – had a rectangle with 70 on it and I hit that with my index finger. The bowl inside spun round and round until the thing went 'beeeeeep, beeeeeep, beeeeeep' and I switched it off; then I covered the bowl with a tea towel and left it on a sandstone slab in the cool of the old farm dairy, stirring at intervals with Malcolm's long winemaking thermometer until the red line dropped to 98.

Up to 70. Down to 98. Clear?

Then I added some freeze-dried 'starter', stirred it all once more, covered it again and moved it to the kitchen, kept to an ambient temperature by the heat of the Rayburn.

By morning, it was a gently wobblesome glob. Not solid, not set, and not fluid either. Somewhere in between, a little like the jelly kids used to eat from square greaseproof paper dishes at parties in the sixties. That sort of consistency. It was then ready to turn into a boiled and sterilised double layer of muslin cloth whose corners were brought together and passed through a loop of string so that it could be hung up from an old meat hook back in the dairy. The whey drained into the bowl placed beneath it while the curds remained in the cloth for salting and flavouring. There you have it: cream cheese.

We experimented with several seasonings for this fine artery-clogging stuff: chives, fresh from the garden, or small pieces of chopped pineapple for an authentically sixties-retro taste, but our clear favourite was the divine mixture my pal Jane dubbed Better Than Boursin, flavoured with garlic, fresh parsley and thyme. It was easily the most versatile, transforming dull crackers into a gourmet experience and soufflés into savoury froth; it was far and away our best-selling product and quite possibly addictive.

On Fridays we churned butter. We decked ourselves out in white nylon coats and carried our vats of heated, cooled, matured, stirred and brought-to-blood-heat cream down to the dairy. The cats hovered hopefully in the doorway as we fired up the churn and scrubbed the butter worker in readiness. We had bought this beast via a radio advertisement from a farm in Patterdale. The tray had been almost reduced to powder by several generations of woodworms, but enough of it remained for Richardsons, coffin makers to Penrithian gentry, to fashion an exact replica. The roller was not past repair, so we repaired it and painted the enamel handle British Racing Green. Just for a laugh, really.

The other Friday task was to visit Maggie, who lived a matter of yards up the road at Hagget House, once the village forge. Maggie liked buttermilk and she hadn't had buttermilk since she was – ooh, much younger. As Maggie was well into her eighties that left plenty of scope but she did love her weekly dose of our butter by-product and often drank it straight from the bottle. Maggie had running cold water but relied on a kettle suspended over the range for the hot sort, no flushing loo just a dig-out khazi, which was dug out on a weekly basis by a long-suffering distant relative, and made few concessions to the fripperies of modern living. As a former village postwoman, to this village and several surrounding ones, it was estimated that Maggie had walked the equivalent of at least a couple of circuits of the world. Unsurprisingly, she suffered fools neither gladly nor at all and took few prisoners – as one mobile butcher found out to his cost.

Maggie was considering the purchase of a joint for the weekend. The exchange between her and the butcher, an accomplished salesman, went something like this:

'Lovely piece of meat, this, Miss Kendall. Do nicely for your Sunday lunch.'

'Turn it over, young man' – he was approaching retirement but to Maggie he was just a young pup – 'and let me see the other side.'

'Lovely joint, Miss Kendall. Very tasty, that'll be, with a few tatties . . .'

'Turn it over.'

'It'll cook beautifully, this, Miss Kendall. Nice cold on Monday, too.'

'Turn it over.'

Realising he was squarely beaten – he should really have known better – he turned it over.

It was then that she looked at him, very, very directly indeed.

Not another word was spoken and he put the piece of beef – topside on one side, rolled trimmings underneath, back into his van and left.

If he had had a tail, it would have been between his legs.

All that seems part of another life. These days the rhythm is different: no cheese making on Wednesday, no butter churning on Friday; only Saturday is different now, as it is the day guests in the cottage change over. By eleven thirty on most Saturdays, I'll have stripped the beds, washed the sheets and hung them out to dry on the line in the yard that doubles neatly enough as a garrotting device for unsuspecting visitors, picked up the lawn clippings, watered the greenhouse and fed the sheep. Today, it's eleven thirty and I've just got up: it pays never to be a slave to routine, you see.

November

There's the faintly goaty smell of tup in the air now, in with the ewes by Bonfire Night. Then it's time to put out bowls of Rice Krispies in the evening. Why? Because the pixies come at night and paint the ewes' bottoms brilliant blue or pillar-box red and everyone knows that pixies like Rice Krispies. That's why they're on the box. Ewes' colourful backsides are nothing whatsoever to do with the raddle-bibs that well-dressed tups wear, you see – it's the pixies and their paintbrushes. Honest.

THE CALENDAR YEAR is an administrative affectation. The livestock year does not work in the same way at all; whereas the calendar suggests that November is nearly the end of the year, in livestock terms the month is the start, of sorts, to a new one. Bonfire Night, you see, marks the beginning of the mating season when the tup goes in with his flock. With a modicum of luck and a fertile tup (I'm almost a poet without quite knowing it, there) lambs will start appearing on April Fool's Day the following year: there's a message in there somewhere.

Power cuts start in November, too. That's because it is now that we really need light. It may be that a bullock has chewed through a vital length of flex – you can easily tell which one is responsible as he is the one with his eyes permanently lit up – or it could be that one of the horses has scratched, itched and rubbed on a pylon and somehow succeeded in blacking out the whole village. It happened once, very embarrassingly. The man from the leccy board came round the week after, with demonic expression and several rolls of barbed wire; Rajah, our retired police horse, looked on with a face rather longer than usual as the engineer fixed it all in place: so much fun, so easily spoiled.

Our Sunday walk through the woods takes us through a cocooning tunnel of russet and gold and bronze trees along the riverside path. Some friends went to New England once, for the fall, but they reckoned they'd wasted their money. 'We should have just set up a tent on Armathwaite Bridge,' they grumbled. 'The colours there are just as good.' And with what they saved on the air fare they could have bought a Secretary Bird sculpture for the garden.

Our walk today ends, like so many favourite country walks, at the pub; the new licensees greet Gyp and Tess, or G&T as they are better known, with a predictable 'Aren't they lovely. Are they related?' Tess gazes up lovingly at her new potential best mate who mis-guidedly asks her: 'What do you want then?' Tess, all doe-eyed innocence and charm incarnate, is far too well mannered to presume on someone's generosity so she doesn't say, 'Well, actually two slices of cold roast beef usually go down rather well . . .' This is, after all, what G&T expect, having established the pattern throughout the previous incumbents' tenancy, but their dumb pleas go unrewarded. We sup our pint, they don't get any roast beef and we all go home.

It must be very confusing being a dog sometimes.

Mickey has learnt an uncomfortable lesson: if you annoy show jumps, they fight back. This is probably just as well, as tying hedgehog skins to show jumps is frowned upon in these enlightened times. He had a bruise on his shoulder and a sorry look in his eyes but a few carrots and soothing words smoothed the situation over and by the weekend he was popping poles again.

Malcolm fell over too but carrots and soothing words failed to work the same magic for him and he ended up going to see the doctor. Unfortunately for Malcolm, the practice dispenser keeps her own horse at the same yard as Mickey and, having heard of Mickey's tumble, she enquired fondly after his welfare. We simpered and twittered over the horse but Malcolm garnered so much less sympathy that he was moved to ask me afterwards: 'Do you think it might help if I grew a tail and a couple more legs?'

I said no, it wouldn't. Just think of the extra cost of resoling all those shoes. Not to mention the number of Blakeys segs.

My mum used to say that as you got older bits fell off. It is difficult to argue with the logic of someone whose one other philosophical pronouncement was 'One half of this world is put in it to aggravate the other half.' She may well have been right on both counts but in my case things aren't so much falling off (yet, anyway) as changing colour. I am morphing from redhead to greyhead and may even have to start cruising the colourant aisles in Boots any time now. It could be age. It could be that I am turning into a squirrel. For now, I am still approximately the colour of a young orang-utan but the grey threads are definitely on the increase. It might be, of course, that keeping animals is enough to make anyone go grey.

Not that the new arrival is causing me sleepless nights. Rather the reverse; she is doing rather well. She is called Blossom. Whether I shall ever live down having a pony called Blossom only time will tell, but Blossom was found wandering on an industrial estate in Newcastle and she wasn't looking for IKEA. She was in a sorry, emaciated condition, reduced to being a mobile home for lice, worms, mites, two different sorts of mange and myriad other parasites, all vile but none so repellent as the creep who allowed her to get into that state. Her immediate future looked bleak.

Blossom was taken into the care of the RSPCA and they set about a comprehensive de-bugging programme worthy of the very spookiest spooks in MI5. The dedicated staff at the rehabilitation centre – it is not a sanctuary or a rest home – first restored her to health and then broke her to ride and to drive, taking the view that ponies of Blossom's age and size ought to have a useful working life rather than stand idly in a field like overgrown garden gnomes. More than a year on, Blossom is, well, blossoming. So much so that when I took her papers to the vet he squeaked: 'Fourteen-two, five hundred and ten kilograms? What is it, Jackie?'

'A vanner. She'll be to lock up during the week of Appleby Horse Fair,' I replied.

Like me, she is a good doer. She lives on fresh air and a nice view and at that impressive 510 kg she is unlikely to trouble the runners in the 2.30 at Cartmel, but I am anxious that, at the earliest possible opportunity, she learns to differentiate between sods and feet. Especially mine.

Blossom is piebald, and where most normal horses have a leg at each corner, Blossom has an extravagantly plumed Corinthian column. All her 'feathering' had to be clipped off to evict the mites that had set up bijoux residences in its darkest depths but now it's thickening up nicely and grows ready-crimped. Atop her short, fat hairy legs is a bottom exactly the shape of a ripe Granny Smith; no 'pear shape' girl, this. Her predominantly white face has far more whisker and beard than is strictly feminine and a permanent 'winsome fruitcake' expression; a pair of big, kind, kooky eyes that have seen the callused underbelly of life look out on her new world, from the hill and across the valley, with a wisdom beyond their tender years.

Unlike many piebalds, Blossom's eyes are edged with black; in a poor light, she looks a bit of a Goth with her startlingly white features and all that eyeliner, but as she's got no obvious body piercings, her legs would not look good in tight black trousers and I can't see her being much of a Siouxsie Sioux fan, I prefer to think of her as more in the mould of a heavily kohled frisky fifties starlet. Think Elizabeth Taylor playing Cleopatra and you're just about there.

If she's La Taylor, little Bob our borrowed Welsh pony is Richard Burton. Within minutes of her arrival, Blossom got down and rolled at his feet so I can only assume that she and Bob were close personal friends in another life, because an act of obvious submission like that is difficult to explain otherwise. There is usually a square-up of some sort when you introduce two horses to one another, a bit of Me-Tarzan-you-Jane chest drumming, and then, once they've worked out a mutually acceptable pecking order, they settle down. Bob and Blossom simply didn't bother; instead, they spent the rest of the day grazing contentedly together, so instead of darting up and down stairs to watch that they weren't trying to kill one another, I went shopping.

The next morning I drew the curtains to see Blossom sound asleep and Bob standing like a small self-appointed sentry guarding her.

The morning after that, they were both flat out and dreaming for all I know. They could even have been snoring.

By Friday, they were sitting like bookends, noses almost touching, each inhaling the other's breath: deeply unhygienic no doubt, but very sweet.

Mutually besotted: a happy state of affairs.

On our first excursion around the local lanes I learnt a little more about Blossom. She doesn't care for drain covers, thinks Beltex sheep talk a foreign language and in all probability has not seen Holstein cows grazing in fields before. As she peered at them intently, a thought balloon popped up between her ears. It read: 'Are we by any chance related?' It's reasonable: their body markings are strikingly similar. Blossom has been extremely well brought up and doesn't hunt in pockets for Polos or nudge for carrots; she takes big deliberate steps and feels rather like a favourite armchair to ride.

I think we shall get along just fine together.

Mickey has come home for an end of season holiday. Mickey is sixteen and a bit hands, chestnut, gorgeous and improbably athletic, and Blossom thinks he is as hunky a chunk of horseflesh as she has ever seen so she mooches over to the fence for a chat. Mickey tries his best to make friends with her and perhaps tell her a few impressive and possibly wildly exaggerated tales of his glamorous competitive lifestyle but Bob, thirteen and a bit hands of jealousy incarnate, will not stand for having his woman stolen. With teeth bared and ears flattened against his chiselled little head, he warns Mickey off. Mickey does not need telling twice but the spectacle of two devastatingly handsome chaps fighting in a gloves-off, no-holds-barred fashion over the attentions of a cob whose ancestors surely pulled carts is one to gladden the hearts of podgy, common girls everywhere.

The RSPCA will keep an eye on Blossom and when the inspector calls he will park his van in our yard. Those unaware of the true purpose of his visit will indulge in some fevered speculation. Have I been

25

loading the mouse traps with bait higher in salt than government recommendations? Have I been verbally abusing the sheep – again? Or inciting G&T to tread on the cracks in the tarmac? Yes, I am confident that the RSPCA van's appearance will fuel the village rumour machine for as much as two full days, if nothing much else – divorces, burglaries, or planning applications – is going on.

You, of course, will know differently.

Trying to move the sheep this morning was a complete disaster. The ewes trotted through the gate obligingly enough although they made it clear they didn't greatly care for the mud; then they took leave of their few remaining senses and careered about the back field instead of obediently following me into the pens. I thought they would be fine, especially as the tups were already penned there. Clearly, they have no desire to get it on with the boys, who, of course, feel rather differently, Senior Boy having spent the past week leering through the fence at them.

Freddie, my chief of staff in all matters ovine, arrives and as we both realise we are on a hiding to nothing we adjourn for coffee. He has his so weak it can hardly crawl out of the cup. A lesser man might have hoped that an hour or so messing about with sheep would provide the perfect excuse to avoid trailing round the outdoor markets with his wife but Freddie is made of sterner stuff and, refreshed by appalling coffee, strides off for Penrith.

Never mind about the sheep, I say, tomorrow's another day.

Freddie arrives next morning at nine thirty on the dot, perkily confident of having to spend at least an hour capturing the sheep, another hour doing whatever it is we need to do and a further half an hour on the post-operative coffee. Not, you understand, that he wants to get out of Christmas shopping at the Metro centre. The sheep, though, are already penned, standing in a neat line with mouths open for their medicine, so he'll have to go and, worse, feign enjoyment or possibly delirium at the prospect of separating himself from large chunks of cash. I tell him that he should have gone to hospital after bumping his head at work; at least then he would have been able to plead insanity. It has always worked for me.

We divide the sheep up successfully and walk the five baby girls up to the top long pasture. The next six, all over the age of consent, go into the middle field with the tup, who curls his top lip back lasciviously at the heady prospect of dawn-to-dusk sex. In his enthusiasm to get going, he leaps on the first and nearest thing available: it is an upturned bucket, weak on foreplay and responsiveness. He bounces on to one of the ewes 'twixt pen and field and she does not look as if she greatly enjoys his advances, much less his performance.

The little tups stay in the back and are beginning to fight for supremacy. But nothing like as much as the big tup, who is fighting everything in sight, me included. He seems to have forgotten that I feed him. And while I'm doing that, I need full body armour. He has already shattered one bucket. Mostly, it is OK if he scores a direct hit, as I just get his skull hammering into my backside, where, it is true, there is ample padding. I usually sit on his head for a minute while we both gather our wits, then I grab his horns, gaucho style, leap off and try to turn him in another direction. Then I run like hell.

If his aim is not absolutely straight, though, bloodshed could ensue . . .

In pursuit of environmentally sustainable solutions to twenty-first-century living, we have been exploring the possibility of solar energy. And yes, there is enough sunshine in Cumbria to produce such a thing. Today, a third engineer came to investigate. We didn't think it was much of a problem, two engineers having come with tape measures and gone away smiling.

All is not as it seems. This guy, the Big Cheese, says the roof is too short, the tank too tall and the airing cupboard too narrow. As one who sees herself as not so much overweight as undertall, I can see where this is going.

'Could the tank go anywhere else?' he asked. Well, it could, possibly, but it might mean dismantling and rebuilding the house and that seems a bit drastic. Not to worry: another feasibility study is scheduled for 'early' tomorrow morning. Then we might have a definitive answer. And then again, we may not.

Briefly, Malcolm considers the possibility of putting the tank in the understairs cupboard. I do not think this is a good idea since the understairs cupboard is a repository for everything that doesn't belong somewhere else. Moreover, a disturbing reproductive process occurs down there. Every so often it is cleared out to the point of tidiness; then the door is shut. Six months later, a whole new pile of detritus has accumulated behind it – old jackets, dust sheets, plastic beakers, dead rugs and broken walking sticks appear. Most of them have been thrown away several times already but they seem to have an incurable homing instinct. Where would they all go if we put a hot water tank in there?

We are going to investigate water power as a source of alternative energy instead. There's enough of that too; some might say there's a surfeit, but that won't stop someone in June next year talking about shortages, hosepipe bans and restrictions on car washing. I don't need the incentive of a restriction to prevent me from washing the car but I'm never having another black one, because they show every speck of dirt and the sets of fingermarks on the boot make it look as if there is a small infant trying to escape from within.

Swapping our fat, gas-guzzling 4WD for this lean little machine caused a flurry of feverish speculation that the folk from Rowfoot had finally gone broke, so modest is our transport now. If the Black Blob had an argument with a tractor it would come a rather poor second and you can't see over the hedges to do your shepherding from its front seats as you could in the 4WD, but those considerations apart, it's a neat little vehicle. If you were absolutely desperate to see over hedges I suppose you could stand up, but you would need to have the model with a sunroof so that you could stick your head through the hole like a sort of human periscope. Otherwise, I wouldn't recommend it.

*   *   *

Vehicles, you see, are a barometer of how well a farmer is doing. Drive a spanking new turbo-charged 4WD with leather upholstery and it will be assumed that you had a record harvest, superb hay crop and 300 per cent lambing, and that your B&B has been full throughout the entire season. If you buy a new tractor – an iridescent orange Fiat Agri or a brilliant blue Ford with matching muckspreader – you can add assumptions of stratospheric milk yield to the litany of high achievements above. It may be, though, that the trip to the Banyan Tree in the Maldives may need to be put on hold in order to finance the credit agreements for all this colourful transport. Of course, you could just head to Southend for your annual holidays or, if things get truly desperate, the next park bench along from where you spent a clement three days last year. If, like us, you drive a Toyota Yaris, don't have a trailer of any hue at all and do everything about the place using a shovel with a woodworm-infested handle – well, you can draw your own conclusions.

Machinery is OK when it works but when it doesn't it involves unnecessary expenditure of time, money and patience: that's why we don't have any. Besides, tractors have minds of their own; they're bigger than me and I prefer not to pick fights with things that possess so obvious an advantage.

I saw one going for an afternoon drive on its own last week, down by the river: not a good location for an unsupervised machine weighing several tons. It reminded me of another, several years ago, that had been similarly trusted to behave impeccably in second gear as the driver ran behind it, scattering the turnips it was chopping so neatly. The tractor hit a stone, turning its wheels sharply right as a result and knocking itself out of gear. Down the hill it went, in neutral and at some speed, breaching the dyke at the bottom and smashing a couple of trees before careering across the road, through another dyke, and into the next field, where it turned a spectacular double somersault and lay upside down with its wheels still spinning and its chopper still chopping.

Its rightful owner was in hospital having some delicate plumbing attended to and finding the one o'clock news quite stressful enough without having his worries compounded by reports of fugitive tractors. So we were all told not, on any account, to mention it to him. The exchanges around his hospital bed went something like this:

'How are you?' (Don't mention the tractor.)

'Eh, lass, not so bad. Be glad to get home.'

'How's the food?' (Anything not to get on to the subject of the tractor.)

'Terrible. Can't even recognise half of it. Some of it's grey.' (The tractor's grey. Strange coincidence.)

'I expect it's chicken.'

'Family all well? Any of them home this weekend?' (Don't for God's sake ask about the tractor.)

I suppose he did find out about it when he got home. But probably not until the insurance claim came through.

One way and another, the gripe and broom cause far less trouble.

It is now sufficiently chilly in the evenings for the Rayburn (a poor man's Aga) and the multi-fuel stove in the sitting room to be lit. To keep both these monsters fed through winter, a heap of logs about the size of a bronze age tumulus is needed. There are a few things you need to know about logs:

- They need to be dry to burn.

- They need to be hardwood, or you will be up and down more often than a tart's knickers feeding the flames.

- Woody the Log Man will bring you good, dry logs in response to your first order. The second order will be the same but smaller and the third order will consist almost entirely of kindling riddled with woodworm and will burn like tinder. You will remonstrate with Woody who will assure you that he brought small logs because he thought you really needed it for your dolls' house and would appreciate making savings on firelighters. You will order once more from him, then change supplier and the whole cycle will start again.

This year, much like every other, I decide to try a new supplier with a website and an email address – the wittily named Fuels 4 Fools, or something very like it – thinking it would constitute a neat fusion of the traditional and the technological.

I really ought to have known better.

Firstly, he rarely checked his emails and then he phoned at about 4.30 in the afternoon to say he was 'on his way'. This, let me remind you, is November. It is twilight by 2.30 and by 4.30 any remaining light is fading fast. I do not know where he left from, but since he, or rather his lackey, arrived at 6.10 I can only assume it was somewhere the other side of the Pennines. Or perhaps south of Manchester. By 6.10 the sky is like pitch; a depth of darkness that serves as a powerful reminder that Cumbria is one of the few places in the United Kingdom that enjoys a proper night sky. You might find this difficult to believe but itinerant astronomers constitute a significant niche tourist market. Itinerant astronomers, though, don't need to unload logs.

I had ordered a winter's worth of logs, a socking great trailer full. They were to go into the End Shed, an exterior version of the understairs cupboard where there was just room for them to be squeezed in between the redundant rusting hayrack and the blue trailer. The driver might have worked out for himself that delivering several tons of logs in the pitch dark was always going to be a challenge, but at Rowfoot the problems were compounded by difficult access as the field gateway is uphill and on a bend. This had all been fully explained to the Gaffer but the Gaffer had inexplicably neglected to mention it to his driver. So the driver ended up worming about a bit on the bend, wellying the accelerator and taking a run at the gateway before becoming predictably and exceedingly stuck. With an unlit trailer-load of logs blocking the road behind him on the apex of the bend, he was an accident about to happen. I threw myself into the path of at least one passing car to stop it in the darkness before lobbing a few choice phrases about criminal negligence in his direction. He would not be deterred and had another go, this time clearing the roadway before sticking fast again.

'I'll just tip them here,' he said, reaching for the Power Take Off drive.

'You will not just tip them there.'

'Right.'

A long, uncomfortable silence ensued.

'I'll help you to barrow some over to the shed.' And that was him, not me, speaking. Mutinous, murderous and silent, I clenched my teeth and fetched the barrow. We barrowed until the load was three-quarters empty and he could manoeuvre the Land Rover and trailer and tip the rest. And then, with stupendous cheek, he trotted off to 'do the paperwork' leaving me to lob the remaining logs into the shed single-handedly.

'Some of these are not logs at all, they are halved trees,' I complained. 'I ordered big logs – twelve-inch logs, I think we agreed – this one's twenty-three.'

'Oh.'

'What will you do about it?'

'Have you got an axe?'

'No.' And if I had, you might be facing a future involving a great deal of hopping. Or worse.

'We'll stop by with a chainsaw, then.'

'Good. Don't forget.'

I think I fell into a bath at about 7.45, cold, exhausted and fuming. At least Malcolm had cooked the dinner by the time I emerged clean, if not actually smiling.

A year on, I am still waiting for the man with the chainsaw to show up. It's not that I haven't reminded them; I have emailed him frequently, but as you know, he doesn't check his emails very often.

Fuels 4 Fools. Oh, how we laughed.

My friend Mary and I decide that it seems a shame not to take advantage of this superb weather and go out for a long ride.

For the past two weeks we have not been able to get into the Coombs, the forest that overlooks a spectacular stretch of the Eden Gorge. The first week, three people had lined up their three cars as neatly as an accounting grid with, maybe, just enough space – if Blossom breathed in and Mary put her feet up on Bob's ears – to get through. Best not. I thought about laminating a sign saying PLEASE DON'T PARK LIKE A PLONKER – LEAVE ROOM FOR PONIES 2 PASS (trying, in vain, to draw something positive, if only witty signage, from my encounter with the Fuels 4 Fools boyo) but I didn't get round to it.

Last week, there was a lively shooting party in the woods; we had taken the precaution of wearing the riding hats without little round circles on the back so that the guns would not be tempted to use us as target practice, but it still seemed better to play safe and stay out.

But today we set off in hope. Blossom and Bob fly up the green lane, sending sods flying. We go gently down the other side, towards Longdales, and come out on to the tarmac lane in time to spy the Kendalls' Christmas dinner standing on a shed with its friend. Two turkeys, looking as if they are working on a script for a remake of *Chicken Run*. They make the most peculiar noise, rather like sea lions with sore throats. At the Coombs this week, hallelujah – no cars, no shooting, just a neat bit of brand new fencing with the gap perfectly accessible. In we go.

In the course of our circuit, we meet one pointer, one Rottweiler, a springer and assorted terriers; we could surely weave these canines into a song if we tried hard and employed some imaginative rhyming. High up in a cloudless sky, a pair of buzzards mew and squeal, and alongside us a red squirrel runs atop the wall, shrinking as murderous gunfire from the opposite bank resonates through the gorge. One of the big estates is having a shoot; it sounds like Fallujah on a busy day.

By the time we reach home, the ponies are sweaty and the light is fading, but at least there are no logs to unload. Just to round the afternoon off, Bob pauses on the concrete path and dumps an impressive mound of dung that sends up a heat plume into the cooling air. It is really quite difficult to know how such a small pony manages such a huge heap.

*　　*　　*

I do love the nightly feeding round at this time of year. Car lights float like little sputniks through the darkening dusk, distant mewling cats and howling hounds scythe through the silence, and the dogs whirl pointlessly about as the tup becomes ever more lethal if he becomes detached from his flock. He has wrecked another bucket and inflicted minor injuries on my shins and bottom in the past few days. I wish he would sort out his anger management issues. Deer have been in the wood, stripping the bark from one or two trees on the boundary, but it is hard to work up much ire where deer are concerned. They are such flimsy, ethereal and delicate creatures that a harsh thought lobbed in their direction might fell them.

It is a complicated ritual, though, at present. The little tups charge into the pens and demolish their feed; the trick is to get the last pen shut before Bob and Blossom arrive through the gate and Bob, who has no need whatsoever of extra rations, ducks in and nicks the lot. With luck, the ponies peel off, Bob to the left and Blossom to the right, without the aid of a headcollar and I shut their doors while they feast on barley, sugar beet and carrots. Then off to see to Mickey; he hesitates as he has to negotiate a small puddle in the gateway. Usually he hops over it and has a buck and a roll in the back field before hoofing it across to a handful of mix in the bowl by the wall. I follow him, collar him and lead him down the yard before going back to feed the ewes and the baby girls in the top pasture and retracing my steps once more.

Blossom, quickly followed by Bob, goes out into the

field again, and then the little tups are released – are you keeping up with all this? Finally, it's time to tuck Mick up for the night, put his bed down and change him into his smart new nightie with leg vents and cross surcingles.

Worse than a tribe of kids, this lot.

December

The season of goodwill. Nuff said. This year, instead of a smoked salmon starter, we shall be eating some cured sheep. It really ought to have a fancy name and, given time, I may think of one. But in truth, it is an elderly ewe's hind leg. I thought it was worth experimenting with, on the basis that if it was a total disaster, then I hadn't risked anything too valuable. Or at least that is what I thought before I ate Manx Mutton, which turned out to be unexpectedly yummy-yummity-yum. And even better cured. I boned the leg, slung it in a Hugh Fearnley-Whittingstall cure, and added two out-of-date cans of Murphys, working on the principle that anything that's good is even better with alcohol. I think I may have discovered the gastronomic holy grail. So if you see a ewe running about on three legs, you'll guess I've run out of cured sheep.

T HE LITTLE EWES are safely in the top pasture, away from the amorous advances of their father. Sadly, sheep have no idea that incest is depraved unless there is a stout, sound wire fence between them to reinforce the message; meanwhile the Boy, who has developed into a stunning, if somewhat terrifying specimen, is getting on with the job in hand vis-à-vis the grown-ups. Blossom and Bob look on voyeuristically, unaware that voyeurism is very nearly as bad as incest. They are thinking of writing a letter to *The Times* and signing it Disgusteds of Tunbridge Wells. *The Times* is safe for now: one or other of them has yet to overcome the inconvenience of being completely illiterate.

The nights now are long and black, the days not much different; so short are the daylight hours that there is little chance to get much done.

Eric, a man so elusive that if there were still space on passports for aliases, his wife would insist on the inclusion of 'The Pimpernel', materialises at irregular intervals. He and brother David are constructing bits of the cottage conservatory and then going away while their labours 'dry', 'set' or 'settle down'. The weather during the past week – horizontal rain and icy blasts of wind that probably haven't paused since leaving Dogger and Finisterre for an awayday in Eden – has meant that we have given thanks for having made an early start with the building. There's a bit of a crisis about the still-wet seam of concrete immediately by the new French doors, but the people coming in at the weekend claimed to be in such need of a break that they didn't care about ongoing building works.

It's not all talk, either. They really don't care. They are every bit as spunky as their mad pet, a little Jack Russell called, with startling originality . . . Jack. Jack is a monster dog in a very small parcel; he has a proper tail, set to 'permanent' on the Waggometer, and beautifully even tan and white markings. Jack darts hither and thither giving every impression of thoroughly enjoying life. He turns out to be yet another rescue dog, further empirical evidence of the mysterious alchemy that brings rescue dogs and their owners to Rowfoot for their holidays with the same sort of certainty as homing pigeons heading for base. And a good thing too.

As soon as the visitors departed Eric and David returned to beaver away in the garth, putting the final touches to that conservatory on the rear elevation (that's planning-speak for 'round t'back') of the cottage. I hadn't bothered greatly about the double glazing nor the poly-carbonate roof but I was very clear on one thing: the roof absolutely had to have twiddly bits. And that's what they were doing, putting twiddly bits on, when I uttered those three inevitable little words that captive builders dread: 'While you're here . . .' Try to keep the jobs sensible, though – checking slates is one thing, deciding that it is time for a new underfloor heating system, harnessing daylight and solar power, is perhaps another. Eric embarked on a short expedition on to the barn roof and found, in no particular order of importance or chronology, a hole in the chimney breast, several slipped slates and one sliding sandstone slab, and there may have been something about a partridge in a pear tree too but I ignored it. With the exception of the partridge, he fixed them all, with the slickness of a computer's Maintenance Wizard. Very clever.

The afternoon's major kerfuffle kicked off as I sat at my own computer in my office trying to book flights to London. And even that needs some explanation – why fly? Isn't it rather hard on the arms? Well, yes, but when I tried to organise a train ticket the man peered at me from behind his vandal-proof screen and said something like: 'A train, madam? On a Sunday? To London? I don't think so.'

'Why not? This is a railway station, is it not?'

'Yes, but there are no trains on Sundays. Not to London. Not direct, anyway.'

When he explained that the only way I was likely to get to the capital on a Sunday was to arise at crack of sparrow's fart, drive to Carlisle, get the bus to Hexham, change on to a local train to Newcastle (are you keeping up with this?) and then wait for an intercity locomotive to King's Cross, my rising temper moved me to wonder just how vandal-proof his screen really was.

'King's Cross? I'm not too thrilled about it either.'

And I left.

So I thought the best way would be to fly and get the tickets on the internet.

On the other side of the wall 'twixt barn and office – it's not really a detached house at all but one slotted in between farm buildings – I could hear a poor approximation of one of Sir Paul's least melodic creations thrumming along in the background. In curious synchronicity with my tapping on the keyboard was the Frog Chorus, rivalled for sheer tuneless repetitiveness only by the dirgiest renditions of 'Mull of Kintyre'. Dum-de-dum, dum-de-dum . . . Must be Malcolm moving bits of wood in the workshop. Can't be. Wrong wall. And since it's December, in the best tradition of pantomime the noise is behind me.

Time to investigate.

The noise had stopped by the time I got up the barn steps and the cause of the noise herself was now grazing peaceably with her usual winsome fruitcake expression. So what had been going on?

Eric had leant his ladder against the barn; the barn is two storeys at the front but built into the hill at the back. It is why we have an eye-level garden and also why whenever double-glazing salesmen ring halfway through *Neighbours* I tell them, with absolute confidence, that there is absolutely no way they can fix the latest triple-glazed, laminate-floored conservatory on to Rowfoot no matter how hard they try.

Eric had climbed the ladder and on to the barn roof to inspect the chimney stack. Fascinated by the unusual spectacle of a man sitting astride the ridge tiles, Blossom had sauntered down the field, limboed under the ladder, curled her nose round the corner of the barn door and very sneakily nudged it open. She peeped inside. Being young and inquisitive and having never taken the trouble to apprise herself of the relevant health and safety legislation she crept in. To the left, some bales of hay and straw, all neatly trussed, and to the right, bags of fleeces. One day, there will be sufficient wool to persuade a mill on a remote Scottish island to weave it into lengths of cloth. Until then, it is stored here.

Blossom, with an alacrity that belies her bulk,

hopped up and had a bit of a rummage amongst the hay and straw. It did not stay neatly trussed for long. Then she padded about in perfect time to the Frog Chorus on the wooden floor – or ceiling, depending on your point of view, fore or aft – blissfully unaware that she was on the precise site of an earlier disaster. Many years ago, a cow fell through the barn floor leaving a large cow-shaped hole in the floorboards and a rather unpleasant conundrum for the farmer to sort out in the shed beneath. That hole convinced us when we arrived at Rowfoot that reflooring the barn was something of a priority.

Eric, a pragmatist by nature, reckoned that it was a good thing Blossom had big feet as they distributed the load – all 510 kg of it – nicely. And how, exactly, did he get her out of the barn? I have absolutely no idea and even less inclination to find out. Some things are better left unexplained.

I did wonder, though, if she had plummeted to an untimely end like the cow before her, how I might have explained it to the RSPCA who had entrusted her to my unswerving care.

Did I say something about the Little Ewes being safe in the top pasture? Not any more they're not. The gate stoop has rotted and tonight all the ewes and the incestuous, conscienceless tup are in there. They all stream out to get to the troughs though, and food beating sex, even the most forbidden sort, every time, I am able to sneak out a hand and grab one of the Boy's antlers.

Having got him, I have to hold on to him, and that proves to be a great deal more challenging.

Partly because it feels safer and partly because I can at least block his vision of an escape route, we proceed down the field backwards as I try to keep hold of him and the bucket at the same time. It is not easy and if aliens were watching they might be forgiven for wondering whether we were auditioning for an inter-species series of *Strictly Come Dancing*. He rears up and threatens me, writhes and extricates himself, knocking me and the bucket flying.

Unusually stupidly, though, he allows himself to be caught a second time, this time as he is eating at the trough. With a savagery of intent that I had previously believed myself incapable of, I upend the bucket on his left antler – why should I care about his dignity? – and hold on tighter. Getting the gate open with one hand is tricky, but I am determined this time. Once through, I release him and he scowls at me.

I ring John, Freddie's son, in the evening to arrange repairs to one gate stoop and a bit of fencing. Next morning, four tups in the back field. Oh joy.

Someone on the radio has just suggested that shepherds invented golf. First, I typed 'gold', but I knew instantly that was wrong because it was the kings not the shepherds who brought gold. Most fellside shepherds round these parts have only a passing knowledge of gold and that's derived from looking in H. Samuel's windows with hopeful girlfriends. No, it was certainly golf. With an f. How did this wondrous theory about golfing's origins come about? Well, the idea was that the shepherds upended their crooks and, just for a lark, tried to hit stones. Yeah, right. Taking a break between the dagging and the dipping, one shepherd cries out to another, 'Go on, see if you can hit yours past that sheep, there, the idle booger lyin' down,' and of course his co-shepherd takes a shot at the stone, lobs it into the ether and clouts the slumbering sheep a crack on the head. If the sheep wasn't already dead, he sure is now.

It's not really a sustainable theory, is it? For a start, shepherds are not like number 9 buses; they rarely travel in pairs, and anyway sheep are not reliably static targets; occasionally they even walk about.

Radio should be more careful. I am sure that dissemination of this sort of frivolous nonsense is not what Mr Marconi ever had in mind.

And that set me thinking about sheep. What do we really know about them?

Well, for a start, they have brains. I know this to be true because I read a report in a newspaper that someone, somewhere in the bowels of a dark laboratory, is doing experiments on sheep's brains. The things some people do for a living . . . You can buy sheep on eBay, Bo Peep lost some and Black Sheep make exceptionally good beer, though whether as ingredient or master brewer I am none too sure.

And that's about it. In the interests of public information, I've been doing some research and here are the results.

Sheep see more than we do. The human field of vision is around 170 degrees: just less than half a circle. A sheep has a radius of 270 degrees – three-quarters of a circle. A sheep needs this extra vision because in the wild it is vulnerable to attack from larger animals that want to eat it. Humans do not share this useful characteristic, because few things want to eat man now that dinosaurs are extinct, lions and tigers don't venture into cities much and cannibalism is frowned upon in these enlightened times.

Each sheep has its own 'flight zone' or personal space, which, if invaded by anything or anyone other than members of its own flock, causes the sheep to flee. If a ewe has lambs, she automatically and by a complicated arithmetical process (I never knew sheep could add up, much less do trigonometry) recalculates that distance. If threatened a sheep will face the intruder and stamp her foot – so, I've noticed, will the average three-year-old.

Sheep are mentioned in the Bible 45 times. Goats get 88 mentions and pigs a measly 2. There is nothing at all about gerbils in the Bible.

Sheep would rather walk uphill than down. This is a problem for Lakeland sheep; some have been marooned on the summit of Skiddaw for generations. Their un-willingness to walk downhill has become a phobia and alleviation of their misery is achieved only by annoy-ing tourists until dusk every day during the holiday season.

The sheep population of New Zealand is 70 million. The human population is 4 million. Considering these figures, it is astonishing that proportional representation continues to flourish in the land of the Long White Cloud.

Australia has 150 million sheep and they all watch *Neighbours*. Only one appears in the programme and it is called Cassie, short for – wait for it – Casserole. Cassie's demise is not imminent but Erinsborough, a nifty anagram (almost) of Neighbours, is a dashed dangerous place to live. People are regularly duffed up, stalked, hospitalised with deadly conditions, given transplants, run over by cars, taken hostage, and burgled. Crime waves like this do not affect sheep; ergo, it is safer to be a sheep than a human if you live on the outskirts of Melbourne.

A sheep can live from 6 to 11 years. The emphasis here is on the word *can* – few bother to survive this long and we beat quite a few of them to it, turning them into roasts and pies before their time is up. Unless they live in Melbourne, obviously.

A sheep has 54 chromosomes (one for every week of the year and a couple of spares).

A sheep's average temperature is 102.5°F, though this increases when they become roast lamb. Hugh Flippin' Eatsitall thinks that the best way of cooking leg of sheep is to bury it in hay. He advises lining a roasting dish with hay, putting the joint on it, topping out with more hay and a few herbs and sealing the lot in tinfoil, taking care not to leave any straggly bits hanging out because they will catch fire, incinerate your house and cremate your gerbil. Insufficiently catastrophic to earn a mention in the Bible, but a reasonable conflagration none the less. Put the sealed roasting dish into a hot oven, get yourself a drink and forget all about it for a couple of hours. Dinner guests will evince satisfying expressions of dismay when they see that they are about to eat a hot bird's nest, but the taste is sublime, reminiscent of haytimes long ago. This recipe has failed me only once and it was nothing to do with the lamb; it had been such a lousy summer that we had no hay sweet enough to use.

A sheep's pulse rate is 75 beats per minute. There is no record of whether, or by how much, this increases at tupping time, lambing time or just simply shepherding time, when the shepherd's heart rate, blood pressure and anxiety levels will rocket even if the sheep's don't.

A sheep has a respiration rate of 16, though if it has pneumonia this can rise dramatically. As well as pneumonia, sheep suffer from a number of colourful diseases: hairy shakes, pulpy kidney (don't put these in your pies), swayback, scrapie, Black Disease (not to be confused with blackleg which is different but every bit as terminal), dysentery, foot rot, strawberry foot rot (rather pinker and tastier than the ordinary sort), louping ill, braxy, rain scab, fleece rot and struck, which is different from strike and more deadly. They fall prey to a multitude of parasites and bacteria, providing endless amusement for parasitologists: another strange career choice. The Parasite Police are an OK bunch, but if they outnumber normal people in your address book in a ratio of three to one then it is time to think about the company you are keeping. Sheep get sunburn, too, so one way and another a sheep is a walking library of pathogens, infections, bacteria and trouble. I'd stay away from them if I were you.

'Making sheep's eyes' is a silly expression. Why would you want to? Except, I suppose, in pursuit of even better vision than mentioned earlier.

This thing about *sheep having brains.* Even though this seems unlikely, DEFRA are conducting surveys of sheep's brains to investigate scrapie, which the panic merchants tell us has similarities to BSE. Scrapie has been around since at least the middle of the eighteenth century so it has taken a while for DEFRA to get worked up about it. If it really had massive public health implications, one might have expected thousands more to have fallen off their perches as a result. While I am at it, it may be worth pointing out that the link between CJD and eating BSE-infected cattle remains stubbornly unproved scientifically. Not that you would ever guess.

Even so, *we are not allowed, in the UK, to eat sheep's heads.* It will not surprise you to learn that this does not apply anywhere else in the world, and there is nothing to suggest that foreigners are dropping like flies from eating sheep's heads. Yet in the UK, whilst we have signally failed to outlaw genetically modified food, pre-sumably because we don't want to tweak the tail of that sacred cow called 'business', we are not allowed to eat things whose deleterious effects are completely un-proved. It's a mad world. Only thee and me are sane, and even thee's looking a bit funny nowadays . . .

Sheep are sheared every year. Some sit still and others wriggle as if they've got ants in their pants. Wool is worth diddly squat but finds its way into clothes, furniture, carpets and house insulation. It is naturally flame retardant – that's why you so rarely see sheep on fire.

The Lord Chancellor sits on the Woolsack. And on his backside, obviously. The tradition of the Woolsack originated in the Middle Ages when wool production was the most important industry in England and the sack itself was filled with the fleece of rare Cotswold sheep, chosen because Herdwick would have given the Lord Chancellor an itchy bottom: it must be an enormous comfort to the Herdwick that he is not responsible for such an unseemly state of affairs.

A final nugget of useless information about sheep comes from my regular correspondent Richard Roscoe, who lives in Armathwaite, a village with lovely riverside walks but no Marks and Spencer, Chinese takeaway or petrol station. Richard tells me that *it is traditional for crofters to dip Harris tweed in sheep's pee to fix the colours.* One can only guess at the aroma that surrounded the crofters after a hard day's dyeing. It is probably safe to assume that Lentheric, when they were choosing a name for an upmarket fragrance, were probably unaware of this fact. Otherwise, why did they settle on Tweed? But thank you for sharing that with me, Richard.

And finally, the noble ovine in verse:

Mary had a little lamb, she tied it to a pylon.
Ten thousand volts went up its bum and turned its
    fleece to nylon.

Unsuitable for the Lord Chancellor's cushion, then?

I should not be dwelling on any of this. I should be preparing for Christmas.

Preparations for the festivities proceed apace. I think this is why the radio bods are all obsessing about shepherds.

Malcolm is in the kitchen chopping things. I am visiting briefly from the building site at the bottom of the yard when a lady from the leccy turns up to persuade us to change supplier and pay less. Our domestic arrangements have her completely banjaxed: 'You're cooking,' she says, looking at Malcolm rolling croquettes like fat little joints.

'So I am,' he agrees.

'And you're outside, building?'

Indeed I am. I'm building two raised terraces, one in either corner of the cottage garden; more precisely, I'm burying two heaps of rubble and tidying up the corners by raising a couple of small terraces in between baking and making Christmas comestibles. It's a modern version of 'The Twelve Days of Christmas': five granary loaves, four sticky toffee puddings, three mince slices, two raised terraces (where did they come from?) and filling one feeder on the pear tree.

She leaves, convinced that we are mad. She may have a point.

Sara phoned today to say she had some news. She's not pregnant; once you've got two kids, she assures me, you've worked out what causes it and you tell your husband to go and dig the allotment. I can't think where she gets this attitude from . . . No, she says that she went to the nursery Christmas party dressed as a reindeer (I imagine it was some sort of fancy dress affair) but she was a bit hazy about what reindeer had in the tail department and settled for a long furry number. This is silly. Reindeer don't have long furry tails. But authenticity, I suppose, matters little to the under-fives.

I can't even remember what the news was now, so obsessed am I with reindeer appendages. I am obviously more stressed than I thought.

A small torrent has been coursing down the road since midsummer. We have reported it to the highways department and nothing much happened. We reported it again and still nothing happened. Today, three days before Christmas, two men in a lorry parked up outside and one started poking the tarmac with a large, thick spear while the other looked on. Then they began digging and wondered whether the leak might include sewage. Then the second fella decided, 'No, no, it's sand . . .' They dug a bit more, leaving a crater in the road with two sandstone monoliths – the old drain covers – piled up alongside, placed red and white planks on stilts round it and Men at Work signs along the road and then disappeared. But not before they had said that they were on annual leave from this afternoon.

The visiting tribe arrives in relays: two on Christmas Eve, four more on Boxing Day and another four the day after. They all claim to be family but I am not entirely sure that I recognise them all. Still, they are here, camped on floors, in layers, in snuggle sacks. There are more bodies scattered about my floors than in a metropolitan mortuary now.

They are wondrously quiet when they are asleep but when they wake up it is a different story. I had forgotten how much noise children make. Is full volume shouting absolutely necessary or is this learnt/deviant behaviour? Do they do it at home? Sometimes, they take a break from trying to kill one another to tell tales:

'Jackieeee, he hit me . . .' Fingers are pointed. Mine, forefinger and middle – think Harvey Smith and you're about there – shamefully, are raised behind my back.

'Jackie . . .'

'I don't want to know. I don't care. Go and play outside.' It is several degrees below zero outside and the rain is lashing at the windows. It is so wet that the sheep have eschewed their tinselly festive raiment in favour of waterwings and flippers and are taking cover under the roadside hedge, peeking out only at intervals.

'Go on, go and play outside. Or shut up.'

Unsurprisingly, the latter option is preferable and peace on earth is restored, but only briefly.

Saturday lunchtime seems the most sensible time to tuck into the turkey, especially as two of the under-fives only like to eat things that are orange, breadcrumbed, processed and deep fried. Generally, they avoid beef, ham ('like *that*' – that is to say ham that looks as if it might have had passing contact with a real live pig, and is not thin enough to read the *Beano* through), or pork, or lamb, though they might consider sausages and those small nuggets of factory waste called turkey dinosaurs. Let's hope those things never walk the earth again.

I put the sprouts on in early October so they are almost done now although no one wants them except Phoebe. I refuse to sit anywhere near her for the rest of the day. Still, there's the fight over the furcula to look forward to. What's a furcula? A wishbone, silly. Enough turkey is consumed to reveal the wishbone, then the kids progress to the serious business of shovelling industrial quantities of crisps and chocolate down their necks. Threats of spots, weight gain, harmful diseases and shortened life spans have no impact at all. I nag them that we're not like our ancestors, who could scoff a full fry-up breakfast every day, drink a bit, smoke roll-ups and finally pop off with a heart attack aged about ninety-seven because they burned it all up with hard physical labour in the fresh air instead of vegging out in front of televisions and PS2s. None of that makes any impression either so I shut up.

I am becoming a boring old fart.

The dogs also consume prodigious quantities of chocolate – they are rather less keen on crisps, though they are not averse to an occasional Twiglet – and they end up with poorly tummies and very glum expressions. Serves them right, but when did you ever hear a dog say, 'Gosh no, that's frightfully nice of you but I simply couldn't eat another double rum truffle with white chocolate coating . . .'

The turkey loses weight faster than a celeb on the Atkins diet. It is very, very good. So 'twixt bird and pudding, we pull crackers and put on stupid hats. Mine doesn't fit. It never does: I've got very thick hair.

It looks as though we have the holistic collection of Christmas crackers. Amongst the 'high quality gifts' are a tape measure (a cruel thing to give anyone at Christmas, I think), a notepad (on which to confess all dietary sins), a padlock (to put on the boxes of chocolate) and a bag of marbles. We were all unsure about these at first; marbles seem a rogue gift, not possessing the same sadistic intent as all the other items. Of course, our imaginations could just have been sent into overdrive by cheap Merlot but finally a persuasive suggestion emerged: marbles are what you lose once you are declared of unsound mind brought about by failed guilt-induced dieting. Then it all fell neatly into place.

The kids despatched to bed, the remaining adults settled down to a bit more Merlot and the usual quotient of dire television. Winner of *100 Top Movies* turned out to be *Grease*, which trounced *West Side Story*, *Mary Poppins* and *Oklahoma* amongst others. It is just uppermost in the public consciousness, Sara thought, because there's a revival on in Manchester. 'You can always tell who's been to see it,' she said, 'because they come out of the Lowry singing and dancing.' Worrying, that. And what would Lowry, painter of chimneys and undernourished northerners, think of such hedonism?

The dogs were on the scrounge again, having forgotten the results of their earlier begging forays. 'If you go on like that,' I said wearily, 'I'll set you both up with a new cardboard box each and send you to Waterloo station.' They slunk off.

It had been a long day.

It's all about kids, Christmas, isn't it? Especially kids like Phoebe.

'What gifts did the three kings bring Jesus?'

There's a lengthy pause for consideration, then: 'Gold, frankincense and myrrh . . . and a desk.'

I think Phoebe wanted a desk for Christmas.

There is the traditional power cut, of course. The lights all go out and the tele dies. A small voice pipes up, 'What do we do now?' 'We talk to one another, like they did in the olden days.' The small voice speaks no more. Its owner is just too shocked.

Bit by bit, body by body, they all left.

The house is wondrously quiet.

Just before Christmas, I had another letter from my penpals at DEFRA. Am I going to keep cattle again? Well, who knows? How long *is* a bit of string? So the way things have ended up, I may, inadvertently of course, have told a porkie. Not a whopper but a lie of proportions such that any plea for clemency would fall on deaf ears at the Ministry where at this time of year they just weave the excess red tape through with a bit of tinsel for Christmas decoration. I said I wasn't going to keep cattle any more. And now I might, after all.

I just like to keep the bureaucrats on their toes.

I like cattle, I really do, but I had developed such an acute case of Form Filler Outer's Elbow that even legalising their presence here was posing insurmountable problems. I think it all started with those charity questionnaires which set out harmlessly enough asking you whether you are 'concerned', 'very concerned' or 'about to enter therapy' over the threat to the Purple Spotted Three Toed Tree Frog once common in Weston-super-Mare and now only found in the upper reaches of the Peruvian Andes. Stealthily the questionnaire goes on to invite your collusion until the final paragraph when it makes a shameless attempt on your pocket with 'OK, punk, how much do you *really* care? £10 worth? £20? Or fill in the sum you think might, just might, salve your conscience, albeit briefly . . .'

At which point you discover the free biro accompanying the form doesn't work and you chuck the lot in the bin. And that probably explains why I said I wasn't planning on reverting to tending cattle in the future.

These days, though, it's not just farmers who are condemned to an Orwellian netherworld of intrusive bureaucracy. Santa suffers too, you know.

Take welfare for a start. Imagine, if you will, Santa perilously balanced on a chimney with the RSPCA Inspector (they don't have Constables in the animal police. I think this must be because they inspect things rather than paint pictures of haywains) positioned below between the rose arch and the midden heap. Santa is offering up silent prayers not entirely unconnected with the midden and the uniformed Obergruppenführer is howling in strangled tones reminiscent of a fell hound in full cry: 'If any cruelty becomes apparent, Mr Claus, we'll bring the full weight of Rolf Harris down on you. You are aware of that? Primetime TV will "out" you. Your Licence to Thrill small children will be revoked.'

Santa replies with an irreverent, 'Ho piggin' ho!'

RSPCA Man is on a mission. 'Have those reindeers [sic] had adequate rest breaks during transportation?'

Santa: 'This is our busiest time of year. They're on triple time, quadruple rations of *Cladonia rangiferina* (that's reindeer fodder to you and me) and the poor blighters' dreams are haunted by the threat of hordes of dissatisfied infants armed with water pistols on the Inglorious Twenty-sixth if they don't come up with the Sony Playstations on time.'

'Emotional torture, then?' The RSPCA man sounds hopeful.

Then, of course, the Man from the Ministry is terribly concerned about passports.

'Passports?' snorts Santa. 'We've got visas for Australia, China and Patagonia too, all correct, all in accordance with new regulations and supplied by the Movement Service from its shiny new offices in Workington, but Vixen doesn't think much of hers. Says it doesn't get her good side . . .'

Ministry Man examines them diligently. 'Where did this one come from, tag number 666?'

'666 – that's Prancer. He's a bit of a devil, Prancer, bred by a farmer in Middle Lapland. Genetically modified, he is. Result of an Embryo Transfer Experiment that, truth to tell, went a bit bandy. His ma was a useless racehorse. At Carlisle last time out she didn't so much come in last as look as if she was setting off early for the next race on the card. His dad was a Highland bull with a personality disorder. A technological miracle, Prancer, even if his lugs do have rather irregular trajectories on account of the double tagging rules.'

The Ministry man is unimpressed as only a Ministry man can be.

'That one looks a bit subdued. Not been administering any Forbidden Substances, have you? Everything properly recorded in your Animal Medicine Book, is it?'

'Oh yes. Nowt forbidden. Not Blitzen. It's Rudolph we have a problem with there. Gets high as a kite on that green ladies' powder from Body Shop he applies to lessen the luminosity of his proboscis.'

When it comes to being quizzed on the fancy new transport regulations Santa starts to lose his cool just a little because he's particularly diligent in that department.

'Club Class without fail. More headroom for the antlers, see,' he insists. 'And when we're travelling we always stop at feeding time at somewhere with a few Michelin stars. We do like our intergalactic food with optimum traction and good cornering properties, me and the team . . .'

When Ministry Man starts on about IACS forms confusion reigns. 'YACS? YACS? These are reindeer, mate, not yaks. And before you ask, my Animal Movement Records are *not* up to date yet and nor would yours be if you had to visit every rug rat, ankle biter and sprog on the planet in one night.'

Like every hard-working farmer, Santa has to bow to the Ministry veterinarian's wishes. When his attention is drawn to Donner's sore foot he has some explaining to do. 'It's on account of her name, like that Donna Karan, see. She will insist on wearing those stupid stilettos even though they play havoc with the chimney linings. I could always recycle her into a kebab, I suppose . . .'

And in the event – though perish the thought – that Cupid or Comet incurred some dreadful injury in the line of duty – an unfortunate entanglement with a satellite dish perhaps – and certain, er, drastic measures had to be taken, Santa could no more tuck into a bit of roast reindeer on the bone than humans were permitted to eat rib of beef for far too long.

That's not an end to it, either. Oh no. In a noble effort to assuage the delicate sensibilities of the hygiene hypochondriacs, the Environmental Health department have inspected Saint Nick's beard for lice, ticks and other uninvited guests and issued reindeer-sized pooper scoopers. The DVLA was adamant about a Public Service Sleigh licence and the police have taken quite a strong line on all those glasses of sherry left out for Santa on his rounds . . . though they have had their work cut out trying to make a charge of 'drunk in charge of nine reindeer, a sleigh and an empty sherry glass' stick.

Back in the real but no less zany world of Guidance Notes for Cattle Passports (page 5 paragraph 2.3, since you ask) the British Cattle Movement Service seeks an answer to the question which has perplexed dairy farmers down the years: 'What is a cattle movement?'

Myself, I always thought it was a cow pat.

And if you think all this nonsense makes those golfing shepherds look positively sensible, then can I submit a plea of insanity? It's hereditary, insanity: you get it from your kids.

January

Brings the snow, makes the feet and fingers glow.
Glowing extremities are fine in January, by the way,
but if they do it in July it probably means you are
radioactive. We don't get that much snow, tucked in
between the Pennines and the Northern Fells, and
in twenty-four years here we have only been snowed
in, properly, twice. On one of these occasions, I
walked to Armathwaite with the dogs and several
chaps in pick-up trucks offered me lifts. Without
exception, though, they said the dogs would have to
go in the truck section, so I kept walking: it's not
that we were ungrateful but my dogs are used to
travelling in style. If 'style' is consistent with a
Toyota Yaris, that is.

T HE NEW YEAR gets off to an unpropitious start when
I find a Christmas tree needle in my belly button.
A New Year's resolution to be more particular about
hoovering is imperative. Will I still be sticking to it in
June? Unlikely.

Worse, it is still raining. It has been raining for
days. The dogs are wet and smelly, the ponies are

wet and smelly, I am wet and smelly.

Mary and I go out for a ride, running between the drips really. I had been trying to make Blossom look vaguely presentable but it was pointless as she spends a great deal of her free time sitting in dirt. She rarely bothers to ensure that the dirt is where her black patches are so she finishes up in need less of grooming than of pressure washing. Blossom is deeply offended by this indictment of her personal hygiene and looks the other way. Bob, of course, is the right colour: he doesn't show the dirt as he is brown already. It is only when the mud dries to a faint taupe crust that it becomes clear that he, too, is a filthy little beggar.

Blossom has committed a second act of trespass – same place, same intent. She has worked out how to undo the latch on the barn door with her teeth and discovered that an entry can be effected with just a little careful yet forceful pushing. I now have to push the inside bolt across and tie the handle on the bolt to the opposite side of the door with lengths of baler twine. It all seems a bit excessive so I have sent for a five-pack of kick bars, two of which we'll use to fettle the barn and another to ornament her stable door, because I don't trust her with that either now.

If Blossom were not quite so highly visible and hefty, she would make a very good sneak thief. What's more she would take Bob along with her and together they would be the Bonnie and Clyde of the equine world. Prior to Blossom's arrival, little innocent Bob exhibited no criminal tendencies, but twice he's followed her into the barn and stolen hay. He's got form now. It is very worrying.

Our elderly neighbour has died and Malcolm is an executor of his will. The old boy said he didn't trust solicitors – roughly translated this meant that he didn't like paying solicitors. Anyway, I told Malcolm and his co-executor that they were barmy taking the job on but they didn't listen.

Now? Well, let's just say I rest my case.

They have to clear the house in preparation for sale and because I am as soft in the heart as I am in the head, I offered to clean out the kitchen cupboards. At least it will prevent the mice currently resident from starting some iniquitous sub-letting racket. I've thrown out anything from Presto supermarket because Presto in Penrith closed nearly ten years ago, so, using incontrovertible logic, anything bought there must be ten years old. At least. I've binned anything priced in shillings and pence on the basis that it is probably past its best. And the less said the better about that stuff festering in tins in the pantry; it may be that it is an embrocation or a cure for something, but whatever it was I think I'd rather have the disease. I didn't open the corned beef (circa 1978) but I never thought wine could go off, although now I'm none too sure.

Malcolm will not be going anywhere for some time, that's obvious. I shall not need to ask the dogs 'Has your daddy had any pretty ladies in to play?' if I've been out alone because their daddy has many happy weeks ahead of him ploughing through a heap of paper roughly the size of Scafell.

When we first moved here in 1982, George and his wife were just the little old couple over the road and we assumed they had been married for ever. They hadn't. They had been married a scant eight years when Madge died in 1983; George had courted her, as the quaint phrase would have it, for thirty years. No simple courtship this, as Madge lived in Sutton Coldfield and he in Sheffield, a geographical inconvenience he overcame by drop-handled racing bike – an impossibly cool conveyance in the 1940s. Even more impressive is the fact that he undertook this cycling marathon every single weekend: greater love and all that, and no doubt by the end of this spell he must have had pretty impressive pecs and thighs.

Some say that Madge's elderly mother disapproved of the match but, of course, the full story – their story and not mine to tell – was far richer and more complicated than that. By the time they eventually married, Madge's mother had indeed died and the happy couple were sixty-five and fifty-nine years of age.

After George's retirement in 1975, they moved to Ainstable. He devoted himself to married life and though theirs was short it was filled with their shared interests of house, garden, walking and countryside. After Madge's death his own zest for life waned; indeed, George himself waned and shrank. Our neighbour Nathan observed with uncanny accuracy that 'He's gan intil less space.' Says it all, doesn't it?

Amongst George's last wishes was that his ashes

should be scattered on Naddle Fell where he had cast
Madge's to the four winds in 1983. He called it the End
of the Rainbow.

But before he reached that stage, the old boy had
become so frail that he looked as if a puff of wind might
send him scudding away on the ether, the only ballast
keeping him earthbound the substantial, hand-built
Crockett and Jones brogues harnessed to his tiny feet. He
had developed bloody-mindedness into an art form and
strategies complex and crafty to conceal his mental and
physical frailties. In his youth – though it was difficult
to believe he ever had one, since he was one of those
people who seemed to have been born with white
whiskers and a bald head – he had honed a fierce instinct
for survival, cheating the elements, death and the tax
man (and jolly good luck to him on the last one, say I)
whilst scampering up mountains in Norway and crossing
the Arctic Circle on skis and photographing high mountains and deep valleys in atmospheric black and white.
Evocative, suffused with that extraordinary light peculiar
to high places, every picture he took he kept: pretty girls
on rocky crags, grass-covered log cabins, sheep and goats
hung with little bells lest they strayed, babbling streams
and rushing falls, mist and cloud in that otherworld
halfway to heaven, verdant lowlands and well-laden
tables surrounded by friends. A catalogue of happiness
and freedom too precious to discard yet filled with such
deeply personal resonance that they were unimportant,
irrelevant even, to anyone but their owner.

His relationship with four-wheeled modes of transport suggested that he should have stuck to cycling but

he continued to drive until well into his eighties – in a pork pie hat, hunched over the steering wheel of his white Ford Fiesta. Another mischievous twinkle, another grimace that may have been a toothless smile, and 'Everything's been replaced on that car bar the number-plate.' It would have been foolish to argue.

In the Fiesta, he travelled to the exotic East: Grimsby. He would arrive at our back door, clutching a list of his planned Grand Tour, complete with names, addresses and phone numbers of ports of call in date order, and say in a shaky voice: 'I'm going to Grimsby.'

Those words filled us with a deep foreboding, especially after the occasion he complained upon his return that whilst travelling down the A1 a lorry driver had drawn level with him and 'did this . . .' He gestured, baffled, with his index and middle finger. Oh dear.

Travelling at 20 mph tops on the A1 in a Fiesta was never going to endear him to lorry drivers, was it now? Especially in a pork pie hat.

Eventually, on an excursion to Penrith, he overtook what he believed to be a line of parked cars but was in fact a line of stationary traffic awaiting a change of lights. The police phoned and asked if we could escort him home. Shortly afterwards, his application to renew his driving licence was refused; considerably dismayed, he stowed the car in the garage but never quite gave up hope, firing it up every Friday afternoon to turn the engine over. Being profoundly deaf George had no idea how fiercely he was revving the engine, or of the fearsome racket issuing from the garage: it all worried Nat over the road, though.

'Someday, he'll fire t'bloody thing ower much and his foot'll slip and whole shebang'll blast through t'garage wall like Chitty bloody Bang Bang . . .'

Quite.

By the time he tumbled base over apex in his own drive at the age of eighty-six after his evening round of feeding the birds, lying there for the duration of the coldest, wettest night of the year, it seemed that his entire life had just led up to this most testing feat of endurance.

He survived, of course.

The words he had half snarled at me many years earlier and with a mischievous twinkle in his eye – 'I'll be bad to kill' – proved prophetic.

Selfishly, I often found him exasperating, difficult and ill-tempered. His deafness had imprisoned him in a silent world and only after his death did I learn just how much birdsong had meant to him; the absence of that exquisite, special music was surely a bitter loss, one that, slewing about and colliding with all the other losses in his life, must have been a final impossible, intolerable, isolating burden. Small wonder he was cranky.

Still, it was mighty hard to feel sympathy for someone who, when complimented on his Victoria plum crop – far too many plums for one small man to consume, or turn or have turned into jam, wine or chutney – bawled at me, 'Do you like plums then?'

'Rather,' I hollered back.

He tottered off into the back kitchen, lifted the lid of the freezer and reached in for a box. 'Here, have these. They're last year's.'

It doesn't happen often but I was lost for words.

The freezer was a further source of anguish. It expired without warning, yet secretively concealed its own demise for several days.

George phoned: 'My freezer's broken.'

'I'll send Malcolm over.'

'You come too. Woman's job.'

Funny, he could hear some things when he wanted to.

I fished out a weighty but flaccid salmon: useless.

'I'll throw this away, George.'

'No.'

'Yes.'

'No, it'll be all right.'

'I'm throwing it away.'

'No. Put it in your freezer for now.'

'No.'

'Yes.'

Silence. His bottom lip and eyelids drooped in perfect unison, very sadly. I thought for one terrible nanosecond he might cry.

Still silence. I picked up a pen and wrote on his clipboard, 'You MUST NOT refreeze food, especially fish. It is dangerous.'

After his death, as we cleared out the utility room, I rediscovered this note written in my best hand, with the words 'It is dangerous' underscored so forcefully that I had gone through the paper.

What of the salmon, then? Carbon dating might have been necessary if its label had not identified it as being frozen in AD 1992. That explained it all, really.

They had been together a very long time, George and the salmon.

I chucked it out.

It is unlikely that he ever really forgave me for this act of treachery. But he surprised me by gifting me a pair of stunning antique Waterford perfume bottles in a Letter of Wishes. I was truly touched.

As an executor of his will, Malcolm had the faintly creepy task of sifting through his personal papers. Amongst the income tax forms from the immediate post-war years, envelopes franked 1948, lists, Christmas cards, empty jotters and desiccated pens were letters in inky copperplate handwriting on tissue-thin paper: letters to his beloved Madge. They are now bundled and tied with a blue ribbon, George's wonderful, lyrical, tender paeans, and I remember him no longer as the little guy over the road but as a man of considerable passion, fire and quite amazing tenacity. George: survivor, adventurer and poet.

He still shouldn't have argued over the salmon.

Mickey's holidays are over and he has gone back to Blackdyke, and to Sarah, who rides him in all his competitions. I know it's inconvenient, having two people with almost the same name in the same book, but that's real life for you, and just to clear up any confusion, Mickey-Sarah owes any fragility in her grasp on reality to a long association with the genus *equus*, whereas Sara (without the 'h') simply blames parenthood. Sarah, then, had borrowed a friend's wagon and loaded Mickey up within seconds. Off they went. But as

Mary and I were heading for the woods, Malcolm had a phone call – hallelujah for mobiles – saying that the lorry had broken down on Armathwaite Bridge. No worries, though; they got back. Eventually. Sarah had to ring the RAC, the RAC had to get Rescue organised and Rescue came in the shape of John from, er, Blackdyke. 'Don't worry. I get a callout fee and a pound a mile,' he said, grinning, as Mickey went down one familiar ramp and up another one.

The postman who had followed them down the hill to Armathwaite had noticed that a back wheel was wobbling like a Rowntree's jelly at a six-year-old's party and flagged them down. It was his good deed for the day.

We may live in the sticks and lack all that metropolitan sophistication but we still need communication. Some might argue we need it more than the cities because of our very geographical isolation. And, in typical 'go-get-it' fashion, we have been trying. First a group was formed (they meet in the pub – now there's a surprise) and initially they considered something called wireless broadband, a system where you have one powerful satellite receiver and others sited to bounce the signals round the village. The scale of this challenge is difficult to overstate; Ainstable derives its name from the Norse word *einstapli* meaning 'village on a hill' and hills there are, spiralling and sloping in every direction. It's a great place to get horses fit in wind and limb and a tough one to bounce satellite signals about. We would have needed to establish a village co-operative to hook up to the internet by satellite and as it would have been a non-profit-making company it was mooted that perhaps we should ask some local farmers for a few useful pointers as they always claim not to be making any money from their businesses either. Why is it that the words 'balloon' and 'lead' immediately spring to mind? My sole contribution to this initiative is the suggestion of an appropriate name: Ainstable Rural Satellite Enterprise . . . work it out for yourself.

But the doubting Thomases said, 'Ooh, satellites and wireless wizardry . . . can we cope with all that? Can we afford it? What about BT?'

And lo, it came to pass that BT set a new trigger point for installing – or at least thinking about installing – conventional broadband and it was a hundred. BT's oversight in appending no obligations to these Expressions of Interest was amply exploited by would-be broadband pioneers Rod and Winston who set off on a pilgrimage round the parish, banging on doors and persuading little old ladies to take time out from knitting mittens to register on the BT website, by proxy if necessary, to say they would be 'interested' in broadband access. Truth to tell, Rod had his doubts about venturing beyond his home hamlet of Ruckcroft (several houses, quite a few bungalows, couple of farms, one pub but it's redundant) to the strange land of Towngate (up the road from Rowfoot, two farms, three cottages, couple of new builds, one barn conversion, no pub at all) in case they thought he was one of the fish men who used to come over from the north-east to case joints on behalf of organised burglars.

This was silly: Rod doesn't look like a burglar. Or a fish man.

And the trigger point was reached. Exceeded, in fact. A massive 108 Expressions of Interest have been registered and now we are digging in for the long haul. What will BT do? Because just as registering implies no commitment, so does setting trigger points.

We are holding our collective breath.

This is the weather I hate. Claggy, everything seen through a veil of pinhead droplets: not grown-up rain, not even drizzle, just damp and drear. The mist hangs above the trees, obscuring the bare upper branches and blurring the distinctions between sky and horizon.

Cattle are damp, their haunches and spines sprinkled with dewy rime; even if the ponies were stabled all day they would not be dry by evening.

The land looks exhausted, dull and fuscous; the vivid greens of spring seem a lifetime away. Yet it can only be weeks.

I have been studiously observing the RSPCA dictum that Blossom must not be fed titbits. If you do that, they said, she will learn to poke around in your pockets, possibly nip and generally become a nuisance with bad manners. And they are quite right.

I have been wasting my time.

Why? Well, today, I took the dogs down to the sewage works for a walk. It was time for a special treat, you see. By chance, I glanced back over my shoulder and there, at the gate on the corner, stood a man and an infant. Together, and with considerable enthusiasm, they were shovelling carrots from a vast brown paper bag – more sack than bag, really – down Blossom's gullet as fast as they would go. Blossom had the temerity to make eye contact with me. There was a different thought balloon above her head today. This one read, very clearly, 'Bugger. I've been rumbled.'

So much for all that about 'no titbits'. She did not even have the grace to look slightly guilty: she was too busy munching. I do not think I will fess up to the RSPCA about this.

In obtaining surreptitious rations of forbidden substances, though, Blossom is keeping up a fine family tradition, started by Granny.

It all began after Granny had broken her third leg. She had been 'discharged' from hospital after the second break; my own suspicion was that she had become so impossible that the nursing staff had simply thrown her out on to the street, lobbing a pair of crutches and a few pithy words after her. But she had fallen again, shattering the first breakage for a second time, and was, probably to the horror of the nurses, readmitted. I don't think they liked the imperious way she clicked her fingers and ordered 'Toast!', '*Sporting Life*!' and 'Wireless!' with absolute certainty that her imperatives would attract an immediate response. And when Granny said wireless, she meant radio, not broadband: just thought I'd clear that up.

Malcolm and I went over to tidy her flat, finding bottles in places you would not believe you could get bottles into: the high-rise cistern in the loo yielded one brandy (Courvoisier, to be specific), one sherry (Fino, Findlaters) and a rather good claret. All empty, of course. The obvious hidey-holes under the bed, behind the fridge and between the casing and the bath were similarly productive, and we didn't even look in the cellar for fear of what we might find.

How, then, did she acquire such a cache, given that she could hardly hobble? A few discreet enquiries of impeccable sources revealed all. Martin, at number 13, was buying a bottle of brandy – Courvoisier – from Stowells on Church Road.

'But she said not to breathe a word to a soul, because, she said, "you know how people *gossip*, dear boy . . ."'

And Mrs Graham, at number 7, was commissioned to acquire a bottle of gin from Budgens, but not to breathe a word to anyone, because 'you know how people *gossip*, my dear'.

Mrs Payne from number 9 had instructions to procure a bottle of 'sherry, Fino – proper Spanish, dear, none of that British filth – but don't breathe a word, because you know how people gossip . . .'

Mrs Hill from number 5 supplied a couple of bottles of claret from the Express Dairy over the road and Mr Bennett from 15 was responsible for the Chablis from Cullens.

A crafty system, maybe, but one of unimpeachable efficacy. So, one way and another, Blossom is just keeping the faith, really. It is entirely possible that Granny is Blossom's guardian angel. She always preferred animals to humans.

*　　　*　　　*

Back here in present day Cumbria, the weather fore-casters have been suggesting Arctic conditions. The media have countered that the weather forecasters are all getting rather overwrought and that it probably won't happen, or it might happen but not be that serious, and that even if it does happen and it is serious, it won't last long.

It's happened.

In a spirit of forward planning I brought the ponies in for the night on Monday and they were quite happy.

That happiness evaporated at about 11.23 this morning. The vet came. Ponies do not like vets because vets smell of the stuff that heralds injections. The only fear I can equate with this is that of the school dentist, although it was always a close call whether the imminent visit of that sadistic personage or of Nitty Nora struck the greater fear into a Mixed Infant. Nitty's zealous metal comb, on the one hand, ploughed tiny furrows trickling with weak, apologetic bloodied rivulets along young scalps. But even that callousness was as nothing compared with the crazed and psy-chopathic activities of the dentist, a woman whose early scientific career began with experiments on fleas. Having pulled all the legs off the flea she had schooled to leap in the air on command, she reputedly concluded that fleas deprived of legs became deaf: a woman, I am sure you will agree, whose loathsome depravity made the Marquis de Sade look like Florence Nightingale.

At least animals do not suffer nit inspectors and dentists, just vets. I am deeply in awe of vets, partly

77

because they deal with such a wide diversity of species from anacondas and Afghans to zebras and zibets (foxy, weaselly creatures with smelly anal glands, since you ask) and partly because they can cure all of them without even being told where it hurts. Although I suppose at least they have the advantage of seeing patients who are neither aggressively litigious nor answer back.

There are equine vets, canine vets, small animal vets, exotic vets (you can spot those a mile off: they're the ones with luminescent bow ties and spats) and I even know of one vet who specialises in sheep, a calling that brings a whole new meaning to the word 'optimism'.

I have met many excellent practitioners down the years and only an odd one lacking in the sort of bed-side manner and cautious respect an old dog or cat deserves. The patients, though, have not always shared my enthusiasm.

Let me tell you about Rajah, an ex-police horse and equine colossus who became a complete wuss with the arrival of a vet.

At the age of twenty-five – that's about eighty in human years – he had been cavorting about doing the hokey-cokey in the top field and broke his pedal bone. 'Box rest,' prescribed our then vet, Tom. Box rest was not Rajah's idea of a good time. Rajah had no wish whatsoever to return to confinement after spending eighteen years in service to Queen and Country stabled, with only the occasional bout of chasing villains, keeping order at football matches (he and his equine colleagues on the constabulary's payroll especially liked to pick up hooligans by their coat collars and dangle them, tantalisingly, just above the ground) and patrolling picket lines to lend a little spice to his daily routine. Once he settled at Rowfoot, he quickly acquired the appearance of an ageing hippy, growing out his hogged mane (clipped down to nothing more than stubble, a legacy of his crowd control days) and re-placing it with the shaggy look, and discovering the pleasures of grass and that undiscovered country, the great outdoors.

It was only under considerable sufferance that he endured six weeks of being banged up before Tom said it was safe to 'extend his area'. It was a nice thought, but a hopelessly impractical one and its merits eluded Rajah completely. We tried, we really did. We let him out into the confined space between the yard and the field where the sheep pens are, but as far as Rajah was concerned that patch was simply his pathway to freedom. The grassy sweetness of the field beckoned irresistibly and there was only a five-bar gate in his way. He worked out that he could do it in, ooh, three strides, and recovering pedal bone notwithstanding – and it certainly would not have withstood much of that – wheeled round, snorted and . . . I launched myself at his head, grabbed the head-collar rope, swung perilously in mid-air for a minute or two in the manner of a footie hooligan and put him back in the stable as soon as the palpitations subsided.

A year or so later, Rajah felt he was due another share of the limelight and developed a nosebleed. Not just any old nosebleed, but one that resisted all attempts to staunch a flow that oscillated between weak trickle and Swithinesque. Our baffled vet decided to 'scope the old boy. Putting an endoscope up the snitch of 17.2 hands of fully conscious hairy monster is, shall we say, a delicate procedure and not one to be lightly undertaken. So, just to be on the safe side, we used a twitch – basically a truncheon with a length of string on the end. A twitch does not hurt, it just stimulates endorphins that becalm the beast. Winding the string round the soft wobbly bit below his nose and above his mouth accentuated the funny whiskery handlebars on his muzzle, and Rajah resembled nothing more than a joke-stereotypical Wing Commander.

Having fine self-control (that day, anyway) I did not laugh and Rajah, sensing the seriousness of the situation, obligingly stood motionless whilst attempting to retain as much dignity as a hirsute horse with an endoscope up his nose possibly could. So preoccupied was the vet with the procedure that he failed to notice that, by a process of judicious moustache-wiggling, Rajah had managed to dislodge the device from his nose entirely and drop it on the floor.

And no, we never did discover what caused the nosebleed. The important thing was that it stopped, but only after ten days.

Kareima, my Arab mare, was not so compliant. Normally good-natured and easy to catch, she sprouted little wings on her heels and went into orbit as soon as she caught a whiff of the vet. It was, I am sure, the smell of Eau de Veterinary Antiseptic that did it: Kareima, not being daft, associated the smell of this fluid with a needle's being plunged into her flesh and hopped it.

Maddeningly, she was willing enough to engage in social chit-chat with vet Paul May, Tom's excellent successor, with the security of a gate between them, but as soon as he and she were on the same side of the thing she behaved rather differently.

During her competitive career, Kareima was stabled every night so capture was never an issue, but in retirement she lived out in the summer. It was just a shame for all of us that her booster jabs fell due in August instead of in some bleak midwinter month, for she caught us totally unawares in that first year of her retirement.

Paul and I walked up to the field, ill-prepared for any shenanigans. I had a headcollar and some tempting morsel – a carrot, or a juicy English Cox's (suitably cored and pipped, natch) – in hand and called her. Kareima whinnied in response – she was always a very vocal horse – and trotted across, floating above the ground, elegance laced with power as usual. She took one sniff of Paul, tossed her aristocratic head in the air, inverted her tail, wheeled round and shot off with a 'can't-catch-me-so-there' glint in her eye. She high-stepped and quickstepped, incorporating elements of the Spanish Riding School airs above the ground routine into an approximation of the Prelim 10 Dressage Test with awesome artistry and technical merit. Paul and I stood there, he with unemployed syringe in hand, me fuming, both of us silently contemplating the origins of the phrase 'high-tailing'. I could have murdered her.

Paul watched serenely as she pranced. Although it took the best part of half an hour, he waited on, patience incarnate. Only a harsh soul would have castigated him if he had taken a certain retaliatory pleasure in the needle-plunging bit on that occasion but of course he didn't.

Since G&T are always violently sick in the car, we usually ask Paul to give them their boosters while he's here, but we used to take Julie the sheepdog – that was her breeding not her occupation, by the way – to the vet for her annual jabs. Now, Julie loved a car ride and would jump gladly into anyone's vehicle, only regretting it when the ride concluded at the vet's. So profound was her dislike of surgeries that before leaving she regarded it as a point of honour to squat obstinately, and in a particularly obscene pose, until she managed to expel a small vengeful turd on to the polished floor.

Sometimes, this took considerable effort.

However, nothing inhibited her keenness for jumping into cars, a habit which once nearly landed her in Wales by mistake. Our departing visitors had left their car door open while they loaded their luggage and Julie must have jumped in and fallen asleep unnoticed on the back seat. Only when the car reached Armathwaite and the occupants stopped for newspapers did she awake and sit up. The driver 'looked in the rear view mirror and thought, "Who do those ears belong to?"' They recognised the stowaway, of course, and brought her home.

We've always had bitches rather than dogs, but now that vets are allowed to advertise it is not beyond imagining that one might do a Two for One offer like Tesco to appeal to owners of errant dogs. It might run something like this: *Dog Owners! Fed up with your dog trying to mate with the vicar's leg? Troubled by Rover's frequent disappearances in pursuit of casual sex? Get him castrated! Pay for the first testicle, get the second one free.* It could catch on.

As we were Direct Sellers of cream and allied dairy products, the perceived possibility of our infecting the British public with hideous disease was high, so the Ministry Vet came once a year to check the cows for brucellosis and tuberculosis. In practice, our cows had a very low incidence of mastitis or anything else that posed a threat to the public beyond that of becoming cholesterol-sodden. And that, like so many things, is all a matter of choice, balanced risk and pleasure. The Ministry Vet decreed that the cows be secured in the byre in readiness for his visit; this nearly always fell in August – I sometimes thought I should have asked him if he would mind injecting a horse and doing the dogs' boosters while he was at it, but I never summoned up the courage. I was young and callow, in those days.

The vet arrived ready-draped in a heavy cotton coat in a colour unaccountably neglected by the fashion industry – stale nicotine yellow – a pair of waterproof leggings extravagantly sprayed with disinfectant and similarly impervious wellies. Thus clad, he would swagger insouciantly into the byre, lift up the tail of the cow at the extreme left and insert a sharp needle on the underside until blood dripped into his waiting syringe. This procedure would be repeated as many times as there were cows – if one coughed, he would step back with noticeable urgency – and then he would load the syringes into the boot of his car and leave.

Quite a full day's work, I'm sure you'll agree. And since our cows constituted a 'herd' even though they numbered only seven or eight, the Ministry Vet would then resume reading *Stamp Collector International* until the next inspection on his hectic schedule.

I suspect that our own vets derived considerable amusement when I scurried into the dispensary one day, yelping for copious quantities of an antidote to the active ingredient in mouse poison. 'Hungry, were you?' one of them enquired drolly. I had to explain that I had caught Lady, our elderly Jersey cow, threading her lengthy tongue through the bars round the water tank and hoovering up the entire contents of a small tinfoil tray filled with little green pellets soaked in Warfarin. We all paid generous tribute to the accuracy and diligence of Lady's manoeuvre and I scuttled off with a tub of Vitamin K and instructions to tie her up more securely in future. Now there's a thought: our cows lived in an old byre, tethered by chains sliding on vertical brackets so that they could stand up, lie down and, I think quite importantly, direct the unmentionable into a gully behind them. So they never had to sit in poo – a great deal more than can be said of some of the modern 'kennel' systems where the muckiest cows are free to wallow should they feel so inclined. And quite a few do.

Tethering is prohibited now. In spite of its hygiene advantages, not to mention the fact that generations of cows have lived like that in byres throughout the world, without any obvious deleterious effects, the EU now decrees that they can't any more. A miserable old cynic like me, of course, questions how many EU Commissioners are able to accurately differentiate between opposite ends of a cow, though there can be little doubt that all of them are well versed in the production of mind-numbing quantities of utterly pointless bullshit.

Another of our Jerseys, Fluff, possessed a very unusual talent: she could grow stones. Usually, I could shift a milk stone, a small pellet of almost solid calcium, by trapping it within the teat canal by pinching the top of the teat between my forefinger and thumb and then squeezing downwards with the remaining fingers to expel the stone. It's a horrible knack to perfect but an essential one if you have a cow like Fluff, and when successful is every bit as satisfying as – and this is a girl thing, I know – popping a zit on your partner's back. The stone fires out with remarkable velocity, and if you are lucky enough to recover it you will find that it is as hard as any pebble found on a beach.

The trouble sets in when it is not successful.

I could feel the stone in Fluff's teat canal; maddeningly, it worked its way up and down like a gobstopper stuck in a hosepipe. As a bung, it was highly effective, stopping up the flow of milk and making mastitis a greater probability with every minute it refused to budge. I tried all the tricks I could think of: warm water to free things up, judicious use of angles as I worked the stubborn little flint up and down . . . nothing worked.

Worried that I might be making things a great deal worse, eventually I surrendered and summoned the vet.

The fellow who arrived was new to the practice, tall, angular and with a face that would turn milk sour at fifty paces, if, of course, we had been able to get any milk flowing. He mumbled and grumbled about its being Saturday evening and said – wait for it – that he would be back on Tuesday if need be. I did not turn cartwheels at hearing this. Instead, I phoned our usual vet at home, on his day off, and, heroically, he came and sorted out both teat and stand-in. The latter was last seen heading down the M6, possibly on Tuesday, never to return. He had the dubious distinction of being the only vet in twenty years who inspired no faith in me whatsoever. And his bedside manner needed work, too.

The Jerseys may have driven us nuts by developing milk stones and stealing mouse poison but at least they rarely required help at calving time. Jerseys, you see, have child-bearing hips. They will even calve to a beefy Belgian Blue bull with no problems, whereas the pure Belgians occasionally need Caesarians to ensure mother and calf both survive. Odd, that, really.

It is *after* calving – or to use the technical if impenetrable term, post-parturition – that you started to hit trouble with Jerseys. They are more susceptible than any other breed of dairy cow to milk fever. Milk fever occurs as a result of the sudden and extreme demand for calcium as soon as her milk begins to flow, and as the poor cow pours all her reserves into her output, so her own system is deprived. Then she falls over. It's not complicated. It really is that simple: she gets down and can't get up again.

There is only one rule with milk fever and it is this: the faster you notice it, the better the chance of getting her back in the upright position sooner rather than later, or worse still not at all. And you do not want the 'not at all' option, believe me, not least because you end up having to deal with Daisy's mortal remains. It used to mean that you had a very big hole to dig and that was bad enough, but now, due to yet another directive from the land of the sprout, you have to arrange for dead cows to be collected. Which is a euphemism for being taken away and rendered and costing you some serious money. The big hole was better.

Nearly all of ours had milk fever at some time or another, so at calving time, next to the bottle of nicely chambréed red on the kitchen worktop was an equally nicely chambréed bottle of calcium. You just had to be careful which one you picked up, though the observant will notice that a cork in the top is a clue . . . Why did it need to be at room temperature? Well, it's a bad enough shock getting a socking great needle plunged into your flesh and half a litre of calcium-rich suspension fed into your body through a flutter valve. The last straw would be if it was freezing cold as well. And that is one job I do not miss: rubbing the cow's hide to disperse the calcium sluicing about in a great subcutaneous puddle.

They never tell you about things like this when you are sitting on the London Underground minding your own business and dreaming of downshifting.

We didn't always manage to get the timing right. Some cows just seem to succumb more suddenly than others, like Petal. Petal did not give any warning, any sign, any indication whatsoever that her system was about to go into freefall. She just keeled over.

One night when I did my late night inspection round, there she was, flat out. Cold ears. Grinding teeth. Distended udder. Milk fever for sure. One bottle of calcium wrought no discernible improvement. Another went in without resistance, just token rumblings from the depths of Petal's innards suggesting that she may have inadvertently eaten something from Burger Master that had disagreed with her. I reminded her quietly that if it wasn't for me pumping her full of minerals she might end up as an ingredient in next week's Burger Master Special and that it would be in her best interests to concentrate on getting her bulk upright. She was unrepentant and unmoved.

In defiance of logic and gravity, I hauled her into a sitting position and propped her up against a straw bale. Petal could not have cared less. She began to shuffle and hope, like the wick of a damp candle, flickered briefly. But she gave up again almost immediately, rather closing down the options. I went off to call the vet but as I tried to shut the stable door, somehow – I cannot begin to explain how – Petal wriggled forward and got stuck in the doorway.

It was four in the morning, it was cold and wet and I had a sick cow wedged in a doorway. I must have been very wicked in another life.

Petal, you will be relieved to learn, recovered, though slowly, ungratefully and very expensively. What's more, she refused to give any reliable assurances about next year.

Pixie also made a respectable attempt to get 'milk fever' entered in the Cause of Death box. She was a funny cow. We bought the slender, fine-coated Pixie in the south of England: she was too delicate a flower to bloom in the colder conditions of Cumbria but I always had a soft spot for her. She was particularly sweet-natured and Bambi-eyed so when she went down (never was a phrase more apposite) with an acute attack of milk fever I was concerned. For maximum dramatic effect, she developed the affliction in the farthest point of the farthest field and once again our own efforts to treat it were insufficient.

The vet – we'll call him Mike for the very good reason that that is what he was christened – having administered all sorts of potions said, 'I'd really like to see her get up.'

You and me both, pal, thought I.

So we settled down to while away the time until Pixie recovered. Mike perched on a tree stump; I alternately prowled about and squatted on my haunches. It was early in the morning and a dewy dawn meant that rising damp in the backside was a very real danger, and that I did not want. I had enough problems with a 'downer cow' who maintained an expression of benign indifference to either our presence or her own plight.

We talked of Gandhi and holidays in the southern hemisphere (merits thereof), of reptile keeping and house prices, of modern-day tractors' colour schemes – red for Masseys, blue for Fords – and the fact that you can get cushions with green John Deeres embroidered on them. We discussed whether space really was the final frontier and how wild the west really was (especially Whitehaven) and whether Queen Bess had been any good at all. We cogitated on matters inconsequential and weighty while Pixie, clad in her protective New Zealand rug, looked across the valley to Saddleback in the far distance and every so often summoned enough energy to flick her sweeping eyelashes in desultory fashion. People went to work in their cars, children went to schools on fleets of buses, weaving along the ribbon of lanes to the towns. Seeds germinated and rabbits stole grass from the fields. Swallows and house martins soared and kestrels stooped on to their prey, and eventually, in her own good time, Pixie, like a latter-day Lazarus, got up and walked off.

Neither Mike nor I had the slightest idea when her enforced inertia ended and the voluntary sort began.

Mike went home for his breakfast and some weeks later we received his bill. It was for no more than the standard call-out fee and the treatment he administered. This was in complete contrast to my last experience in the south of England, when a vet with a double-barrelled name rushed off as soon as he decently could and billed me handsomely for every millisecond spent treating and travelling. I think there was even VAT on the good vibes he sent in my direction during his homeward dash.

We rarely kept beef calves long enough for them to have the opportunity to contract milk fever and anyway, their preference was for a really spectacular dose of pneumonia. These episodes always involved fabulously stringy bogeys, fearful bouts of asthmatic coughing and facial expressions that make basset hounds look ecstatic by comparison. A calf with pneumonia will convince you that the Black Death was no more than an exaggerated hiccough.

Alongside the bought-in commercial bull calves for beef, I decided to rear a few crossbred Jersey heifers. It was an initiative I referred to as Rowfoot Money Making Scheme number 53 (B) and it was a failure. A particularly dismal one.

Crossing Jerseys with fashionable continental bulls resulted in calves suitable for rearing on for beef. Most of the time, this was highly successful, and our Belgian Blue crosses and earlier Limousins achieved good returns. The bull calves unfailingly sold well, the heifers less so. Once or twice we were so disappointed with the price of the Limmy/Jersey heifers, we brought them home and put them in the freezer: a bit of a waste, really, of a potentially good breeding animal. So, I thought, why not keep the heifers, rear them on and put them in calf to a different bull – obviously, we didn't want anything inbred. There's enough of that already, in every species, humans included. The idea was that three-quarter beef animals, with a touch of Jersey milkiness about the mothers, should result in some pretty impressive beasts. And so it came to pass. Until Poppy.

Poppy was badness on legs. She killed – or, to take the charitable view, lost – her first calf and so we had to buy one to 'set on'. Setting on a calf is not too complex a process usually: it can take a little while, but with time, patience and just a little cunning it usually comes good.

We bought a very smart little black calf from Hughie

91

up the road, daubed it in the bodily juices of the dead calf, tied Poppy up and held the calf to drink. We persisted in this entirely fruitless endeavour for six weeks. At the end of that time, confident that Poppy had been well and truly won round, I untied her with the intention of loose housing her for a few days with the calf before putting them out in the field. If only things had been that simple. First Poppy made a pretty good attempt to kill the calf and then decided that the next best option was to absent herself from an unwelcome situation. From a standing start, she jumped the pen gate, gently brushing my husband's ear with her right horn as she did so.

Poppy went to the cast market two days later and I reared the calf on a teat. And strictly speaking, since I was telling you tales about vets, I should not have recounted this tale here as no vet was involved. If bovine psychiatrists existed, of course, it would have been a very different matter since a whole raft of experts could have been employed exploring Poppy's multiple personality disorders.

Finally, of course, there's sheep. No-one with even the most tenuous grasp on sanity calls a vet out to a sheep. You take the sheep to the vet – it makes more financial sense that way. And it will die anyway. They always do, sheep.

February

You know that painting by the great Victorian artist Benjamin Williams Leader, February Fill-Dyke? Well, that's how it is in real life too. Wet, bleak and cold. The Romans didn't even have a name for it until some bloke succeeded Romulus as king of Rome and added January and February to the existing ten named months. Everything looks bare, stripped and dead, except those miraculous drifts of brave snowdrops, defying the general air of despondency and gloom. And in that secret world underground, preparations are being made. We, up here, just have to be patient, hope for a Valentine card on the 14th and eat our pancakes.

THIS MORNING I witnessed a hideous and brutal murder.

Half dressed and half awake – I am not a morning person – I looked out of the window to see one of Freddie's Texel tups launching a vicious attack on one of his mates for no apparent reason. The aggressor charged full tilt head down at his victim, taking him off his feet. Reeling from the blow, the wounded tup attempted to stand; time and again, as he struggled to regain some equilibrium, he was subjected to yet another battery. Finally, and mortally wounded, the prostrate sheep only twitched in retaliation.

His attacker stalked off, merciless in victory. The injured sheep needed a vet to bring his suffering to an end.

I have never seen anything quite like it before and I hope I never shall again.

The rain has been relentless these last few days. The Eden is swollen; currents whip and race towards the Solway and only kamikaze canoeists would venture on to the rapids by the mill at Armathwaite as they are in a permanent state of frothing frenzy.

The ponies, though, are unworried by torrential downpours, horizontal lashings of hail, force nine gales and sub-zero temperatures. Why? Because they are not out in any of it, that's why. They may be hairy and hardy but I still cannot bear to see them huddled behind the barn, backs up and heads down, and anyway, Blossom threatened to text the RSPCA and grass me up if I didn't provide warm deep beds of straw and buckets of beet pulp and barley pretty sharpish.

I reminded Blossom that according to her adoption report she 'lived out throughout the winter and received no supplementary feed, just good hay'. Blossom replied – quite eloquently for a pony – that this was all a wicked lie and that actually she had had a deep straw bed and a frilly negligee supplied every night. We reached an amicable compromise and now she and Bob form an orderly queue at about four thirty each afternoon, earlier if it's wet, and in they come, without the need for buckets, headcollars or bribes of any sort.

It is too wet and miserable to go out riding, so I settle down with the local paper by the fire. First, though, I tidy the hearth with the curiously named companion set – that sort of pyromaniac's cruet that sits by the grate – and manage to set the brush alight.

'How on earth did you do that?' enquired Malcolm helpfully.

'It was quite easy, really . . .'

It's not my day, is it?

Still, the local paper cheers me up no end. As a former member of the Tufty Club (you won't remember this unless you are about fifty – it came before the Green Cross code and the idea was to encourage road safety amongst infants) I am immensely gruntled to see that red squirrel havens are being set up. A gamekeeper is quoted as saying that English Nature is not doing nearly enough to save the reds from extinction. What he really means is they should be using not just 'resources and co-ordination' (benign enough commodities, I guess) but bullets and firing squads to exterminate the invading greys. And that's the bit that the *Cumberland and Westmorland Herald* reports English Nature to be wrestling with, the idea of wholesale execution of grey squirrels. Mostly, local wrestling involves men dancing about in richly coloured velvet underpants and white thermal longjohns, but I expect the fellows from English Nature wear tweeds and sensible shoes.

On the letters page a fellow from London is banging

on about poxy viruses, shilling bounties on squirrels' tails (as opposed to their heads) and, needless to say, nuts. He should get out more.

But in a week when a German was convicted of cannibalism (was his victim tender? What did he serve as a condiment?) here's my suggestion: shoot the greys and eat them. Squirrel is good roasted, braised, casseroled, stewed, fried or barbecued, but take care to marinate first because you can never be quite sure how old the little varmints are.

Enjoy!

Depressingly, elsewhere in the paper it is reported that Eden District Council is to reduce the council tax discount on second homes to 10 per cent. As the apparatchiks in the finance department regard the Old Dairy Cottage – our nod in the direction of diversification – as a second home, we will now have to pay 90 per cent council tax on it. Now, I'm not usually hard to please but even I want to go somewhere other than the bottom of my own yard on holiday, so I have a bit of a problem with this 'second home' malarkey. Furthermore, it seems a bit cruel to penalise us for having invested our own money and initiative in something that brings visitors to the area, especially ones who spend their pocket money liberally while they are here.

Sometimes, I think I really should buy that Hebride and sit on it. Shove up a wall and shoot all comers. I could even get in some early target practice on marauding grey squirrels.

I cheered up, though, when I found a picture on

page ten of the *Herald* of a Bluefaced Leicester sheep with a most improbable mullet, and by page eleven I had achieved a state of heightened delirium while reading the report of Penrith Agricultural Society's annual meeting. The secretary mixed her metaphors, moaned about DEFRA (well, we all do that) and whined about the chaos in the cattle classes (not enough stewards, apparently), commended the numerical strength of the rabbit entries and noted that sheepdog handler Katy Cropper, of television fame, had given a 'more professional' performance than on her first visit.

I bet Ms Cropper was delighted to read that.

I do love our local paper. Only recently has it stopped listing local divorces and names of people who have been prosecuted for not having a television licence, omissions that ensure that the defaulters in society sleep more soundly in their beds at night. My suggestion that the vacant column space be usefully employed as a Who's Doing What With Whom and How Often column has, so far, failed to elicit a response from the editor.

The rain has stopped at last. Today had a springlike feel, a high, bright sun and the land suddenly looking green and hopeful. I thought I saw a swallow but I must have been getting carried away because it turned out to be a leaf: it may be time to book an eye test.

The sheep are keen to come to the trough, though every so often they peer into the top meadow where the grass is lush. Freddie's son John has refenced the gap with fresh planks – a quantity of wooden material sufficient to recreate the whole cast of *Crossroads* – and repaired the stoop. I have told them they will be admitted to pastures new only after they produce something worthwhile: a lamb, ideally.

It's a quiet time for the day job, just now. Feeding, shepherding, running between the raindrops: not much to it.

If the days are quiet, and your evenings need spicing up, you can always eat out. After all, Cumbria is practically the birthplace of the country house hotel. Sharrow Bay on the shores of Ullswater is the original, the blueprint on which all others are modelled. At Sharrow, as it is affectionately known locally, the tucker is astonishing, the view amazing and the service discreet. It remains a unique and special place, attracting the great, the good and the showbiz. The trouble is that it has also spawned a raft of imitations that claim to be exclusive oases of seclusion and culinary inventiveness but turn out to be skewed and comic pastiches of an inimitable classic.

Dingly Dell is one such example. First built as a Gentleman's Residence, the house occupies a favoured site, presiding over the village where the plebs live; wooded grounds surround it, its slightly spurious crenellations enveloped in an elaborate fretwork canopy of tipsily swaying old branches.

The Gentleman runs off with the parlourmaid and Dingly Dell falls into disrepair; trees fall down, tiles on the roof fall off and quite a few of the crenellations hover on the brink, threatening to decapitate local children who play hide-and-seek in the grounds. Some view this as an incipient disaster, others as a public service. Biding their time until a fiscally propitious window opens, a consortium of wealthy local businessmen rescue the property, in anticipation of creating their own sort of haven: somewhere safe from the prying eyes of wives, mistresses and taxmen.

Lubricated by liberal doses of claret and single malt, chaps do deals across tables groaning under the weight of game pies and beef ribs; cigar barons on Cuba post record profits and for a time all is well at the Gentlemen's Club. The businessmen find it difficult to keep staff and dislike washing up – that's what the wives are for, for heaven's sake – but the idea of importing wives, even for domestic purposes, is altogether too preposterous to consider, so they employ resident managers instead.

This couple preside over the Club, jealously guarding their positions as the gatekeepers of paradise. Living in such close proximity to the fire of power, they become warmed by it. But the Gentlemen die as Gentlemen do, partly because of the claret and the single malt and the pies and the beef, and the Club falls into a decline. The gatekeepers of paradise live in faded splendour for a while, in a cottage on a hill, with the benefit of all mod cons and running water. Without a tap. But they tire of their watery cottage and decamp to a Polynesian island, where they live surrounded by girls in grass skirts. They exist mostly on gin.

The hotel passes into new ownership once again, inspiring a new wave of fanciful dreams. The new incumbents are a man who wears fishnet tights – an unusual, even peculiar choice of legwear in mid-Lakeland – and his wife, an expert in quality control. Of the bar stock, mostly. This affects her driving, and her frequent collisions with trees, signposts and, more regularly, verges, lead cruel observers to suggest that she may have a new job as a professional decoy, luring the good policemen away from genuine drunks. Another ignominious demise predictably ensues. And another. Then another.

By the mid 1980s the hotel has come to terms with the fact that it will never feature in the premier league of country house hotels but at least it satisfies local appetites for big steaks and big chips on big plates, except on Sundays when it's lunch in the day and an all-you-can-eat buffet in the evening. In the mid-eighties, the women sit in a corner and the men stand at the bar. It's the law.

The hotel books attest to a moderately successful operation, so the incumbents, tired of grilling steak every night but Sunday, decide to sell. Here come likely buyers: peseta-rich from the sale of their all-day-breakfast operation on an unfashionable Costa, back in Blighty they spy a handsome Gentleman's Residence, occupying a favoured site, presiding over the village where the plebs live . . . we've done this bit, haven't we? They can see themselves living here, in this grand house, in spite of having not the foggiest idea of how to pull a pint or grill a steak. They buy it. They can learn, can't they? And they can employ people. He wears a fixed grin and lots of gold jewellery. She keeps chinchillas. They don't learn. They employ the wrong people.

They quarrel, and the lady and the chinchillas go off to Wolverhampton to live a better life. This is probably just as well because Health and Safety can be picky about where chinchillas live and Wolverhampton is preferable to a hotel bar.

Dingly Dell is back on the market again. Its new and hopeful proprietors depart before Michaelmas without leaving so much as a culinary footprint in the Cumbrian soil and its fortunes spiral ever downwards; changes become increasingly rapid, increasingly incredible. Now it's strippers on Tuesdays, singles nights on Thursdays, but predictably, in Lakeland, both 'business initiatives' are doomed. The strippers get frostbite and the singles weren't really singles at all, just opportunists on a night out. The final lunatic pairing seals the fate of the erstwhile Gentleman's Residence and it is now a Rest Home for the Terminally Bewildered.

There's an elderly redhead attempting a wheelie in her motorised scooter in the grounds; she's wearing purple socks and she has a bottle of Limoncello connected to an intravenous drip. She is telling anyone who will listen that she has never really understood why we rely on the RAC and the AA to dish out the foodie gongs anyway. 'Why,' she's arguing, 'should we expect companies preoccupied with horseless carriages to be the national arbiters of gastronomy? We don't ask Cadbury's to adjust our spigots, do we?'

I expect you recognise her.

Saturday, and it's a fresh day, breezy, small high clouds chasing across a clear sky backlit by mellow sun.

An ideal day for a ride through the woods.

Lofty, leafless branches claw into a marbled sky, weaving an intricate tracery like thread veins on a hag's cheek; the old bracken will soon disappear, fresh growth overtaking gold-dry autumn remains with a suddenness that always surprises. The ponies blast up the back lane, full of energy after three weeks of being rained off and rueful as they gain the summit, as if they had forgotten just how far and how steep is the sharp ascent. They steady after this early, mad burst, and settle; the unaccustomed warmth of spring and frantic activity brings them out in mucky sweat, dried to hard spikes by the time we reach home. It is still too cold to bath them.

At home, there has been a robbery. Where there were two fat monkfish fillets, there is now one – and two very guilty dogs. Neither confesses. Malcolm is apoplectic as the monkfish was a special dish for a special occasion: Valentine's night dinner. It was also a special offer: half price in Tesco. Now it is just half quantity, lending the whole affair a slightly surreal symmetry.

'Smell their breath, see which one of them stinks of fish.'

I cannot imagine why I acceded to this extraordinary request but I did: no joy. I fear the crime has been committed some time earlier but one thing is sure: whichever one of them stole the fish, it's a safe bet she wasn't big-hearted enough to share it with her sister.

Malcolm shrugs. 'I suppose it's my fault for leaving it within reach . . .' Twenty-nine years of living with me has taught him that I am unlikely to blame the dogs before him, especially as it was he who left it within reach of doggy tongues that operate with the velocity and reach of an iguana.

I tell him that fish is good for the brain, so we can expect Gyp or Tess to become proficient at quadratic equations in the very near future. Malcolm is unconvinced. 'They didn't even send me a bloody Valentine,' he says miserably. We have some very nice Parmesan potatoes to flesh out the fish as it were, and we say no more about it.

Collies have a thieving gene, you see, just as surely as they have a herding one. Julie, our first collie, was a mistress criminal with a penchant for chocolate and venison, and she passed on every one of her dubious talents to her daughter Bess. They were partners in the crime of the great Chocolate Heist that occurred one Christmas. Eldest daughter Angela had left her vast Christmas bar of Cadbury's Dairy Milk on a low table and mysteriously, although the contents disappeared, a few shredded remains of blue and silver paper remained, testimony to theft by dog or puppy unknown. Angela was livid and vowed that punishment would be meted out to the culprit.

'How will you know which one stole it?'

Silly question, really.

'I'll examine their poo – the one who poos purple foil stole the chocolate,' she exploded with ineluctable logic.

Why didn't I think of that?

Oh, and who did it? They both did.

Julie flew solo with the venison though. Perhaps she

103

just felt that in amongst all the visitors she was receiving insufficient attention, because, mid-evening, a soft thud came from the dining room. None of us took very much notice: Morecambe and Wise were doing their Christmas Special, a sacred ritual that nothing was allowed to interrupt. A little later, during the commercial break, Julie emerged from the dining room, licking her lips, and settled down to watch part two with a contented sigh and a distended tummy.

The haunch bone lay denuded on the carpet. We had tinned tuna for supper that night.

Tuna was usually the staple that came to the rescue. When we lived in Walton-on-Thames and commuted to work in London, Malcolm used to ring to say which train he was catching. And he would usually ask 'What's for supper?' to establish whether the homeward journey was going to be worthwhile.

One night, I was about to respond 'Liver and bacon' when Julie appeared in the doorway, her whiskers dusted with flour and a droplet of blood balanced on the end of her nose like a miniature ruby. Quick as a flash, I amended the reply: 'Li— Tuna and spinach.'

She also stole lambskin slippers, having a particular and inconvenient liking for the left foot. If only she had pinched alternate feet, life could have been so much simpler.

Gyp and Tess have not had a good week, all in all. Tess has had cystitis and ended up at the vet's. She has condescended to swallow the medication but only if it is buried in some soft, ripe Blue Garstang cheese. I have to get a urine sample from her when the pills are finished.

The vet says that it is best to do this with a saucer but I think that a soup ladle could be easier . . .

And on the subject of dogs, I have a theory about their lifespan. Dogs live precisely long enough to ensure that by the time they die you have forgotten how absolutely impossible they were as puppies. That's why you get another one.

It is Wednesday and Malcolm has an Old Boys' Lunch at Barrow. We set off early, down possibly the loveliest stretch of motorway in the land, skimming the boundary of the Lake District and turning on to the Furness peninsula. The silvery flats of Morecambe Bay shimmer under a low sun and the tops of the Coniston range, jagged and misty-dark, are etched on a cobalt sky.

We stop at Airey's, butchers at Ayside, for home-cured bacon (bad for you, but you only live once), black pudding (ditto) and Cheviot chops. Airey's is one of a small but growing number of butchers throughout the UK that markets rare breed meat; I commend them, others like them and Farmers' Markets to all enthusiastic carnivores.

Amongst the stalls at our local Market is one whose existence owes much to supermarket power. Odd, but let me explain: the Thompsons of Bromley Green Farm near Appleby had been breeding fine Aberdeen Angus cattle and selling them, lucratively enough, to a major quality supermarket. But when said quality super-market sent an impersonal directive indicating that they required all stock to be consigned to Oxford in future, in pursuit of that marketing holy grail, centralisation,

Ted Thompson stopped just short of sending them an equally impersonal directive by return, suggesting something along the lines of 'Get thee to a taxidermist.' He stuffed them anyway, by tearing up his contract. 'We take care of our animals. We treat them with respect and pride and there was no way that their last day on earth was going to be spent travelling on a lorry,' explained Ted. So the Thompsons started selling direct to the public from a farm shop and setting up their stall at Farmers' Markets. They have never looked back.

However impressive the claim 'Went to Oxford' is on a potential employee's CV, it is a quite unnecessary qualification for an Aberdeen Angus bullock. Aberdeen Angus bullocks have no need to saunter the quad, take sherry with the prof and gaze at dreaming spires; they just need to cosy up to Yorkshire pudding.

'Support Rare Breeds: Eat Them!' might seem an odd slogan but it is at least a punchy one. OK, it lacks the sheer silliness of 'My Other Car's a Porsche' in the back window of a Reliant Robin or the arresting accuracy of 'Give Blood – Play Rugby', but it's less daft than it might appear at face value. Each breed, you see, has a surplus for all sorts of different reasons. Take three of my lambs from a few years back: one had a topknot, which is about as bad as it can get for a Manx Loghtan. Breed purists would as soon have a two-headed freak with three green toes on each hoof in the hallowed pages of the Flock Register. Another had paler circles of fur round his eyes than he should have had: spectacles, they're called, and they too are frowned on. Maybe not quite as damning as a white topknot but still a no-no.

And number three: well, his important little places failed to descend and their resolute failure to submit to the laws of gravity meant that he was never even going to get to the starting line in the breeding stakes.

So what did the future hold for these three little rejects? The pot, that's what. They might not have been prime examples of Manx sheep but they sure were good enough to eat. But because rare breeds are just that – rare – you cannot trot into your local supermarket and pick a joint of Manx out of the chill cabinet. Not yet, anyway. You have to find a specialist rare breed butcher and once you have done that, I promise, you will never, ever want to eat 'commercial' meat again.

Airey's have been carrying on their business on the Furness peninsula for a couple of hundred years, so they know a bit about what they are doing. Ignore all those jokes about the A590 being the longest cul de sac in the United Kingdom; even if it were true the excellence of Airey's fare alone justifies a jaunt along its length.

They are 'cradle to ladle' butchers; in that Oldeee Worldeee England that used to exist before flushing khazi-whazies, Airey's would have described themselves as butcher-graziers. They farm non-intensively and feed the stock traditionally, largely because these breeds do not thrive on nitrogen-saturated bluegrass and are pretty unrewarding where huge quantities of concentrates are concerned. They do better on mean-street pastures but they need longer to grow.

You are quite likely to find one of the Brothers Airey changing hats, and just about everything else, as they return from the hill and undergo the metamorphosis from Farmer Airey to Butcher Airey. The animals reared out on the hill are despatched without the need for long journeys and absurd food miles on their clocks as they have their own slaughterhouse. What's more, these guys are all in favour of bringing back hanging – for their meat anyway. I recall being introduced to a Galloway heifer on the hook and being told, tantalisingly, that I couldn't have any for three weeks at least, after which she would be delicious. Like any good woman she was worth waiting for.

Airey's beef all comes from traditional and rare breeds, such as Belted Galloways and strawberry pink Shorthorns, yielding meat marbled with fat. Lamb might be primitive Hebridean or North Ronaldsay, so low in cholesterol that you can indulge relatively guiltlessly. Pork with crunchy crackling and bacon with taste and succulence comes from the likes of Saddlebacks, Old Spots and Berkshires. They might look a bit hippified and hirsute alongside clinically clean, intensively reared porkers but at least these have done proper piggy things rather than being reared in laboratory conditions more suited to something grown in a petri dish than to something with a life, a soul and a brain.

The Aireys make Cumberland sausage on the premises without the chemical enhancement beloved of Eastern European shot-putters and the bacon is dry-cured in salt and goes crispy within seconds of cooking – remember that? Every part of the pig is used bar the squeal and that probably finds its way into the tractor brakes.

The prospect of the bacon, crisp and flavoursome, on my plate tonight, accompanied by eggs from hens so free range that you have to watch out that you don't run them over makes my mouth water.

Greed: it's a terrible thing.

All this talk of bacon reminds me of pigs. Our first ones came from Lesley and Ian at Skirwith. That's pronounced Skirr'uth, just in case you were wondering. Anyone who calls it Scurr-With is considered to be an alien.

Ian and Lesley, our original pig people, all too soon gave up their pigs and turned to farming tourists instead. Ian, whose maxim for a happy existence was and remains 'If you can't fight, wear a big hat', had his reservations about this change of course – he loved his pigs – but love, even of the piggiest sort, is an unstable currency. So, the pigs were evicted, their living quarters turned into beautiful cottages and we had to look elsewhere for good, old-fashioned, traditional swine.

It was then that Edward Brown, auctioneer at Carlisle, introduced us to Ivan up at Roadhead. 'You'll like Ivan,' I remember Edward saying. 'You'll get along well.' I sometimes think there was a hidden agenda, a dark and covert message in there somewhere. Edward probably thought we were as potty as each other and he was quite probably right.

Roadhead is so remote that Ivan always welcomed visitors rather as one might have expected Livingstone to greet Stanley in the jungle, only with more bells on. We came to regard a visit to Ivan's as a day out rather than a business trip since he liked what my mother would have called a Change of Boats (Boat Races = Faces) and never hastened our departure.

The land up at Roadhead is wild, with wide open sweeps of fell shot through by brackish pin-thin streams and patches of gorse, interspersed with rocky outcrops. On a sunny day, the majestic expansiveness of its very desolation takes on a rare beauty; on a grey day with heavy clouds squatting thunderously on the tops, it is a bleak and hostile place. When the wind blows at Road-head, it tips birds from their perches and capsizes anything taller than a terrier; it slicks grass down like a Brylcreemed Teddy boy's DA, forces trees to bow and scrape with its tearing, wrenching gusts, and whips the water in field troughs into seaside ripples and waves.

The route to Ivan's farm involved miles of narrow lanes that unravelled as randomly as skeins of wool, winding through tiny hamlets and scattered farms and then – it was a sort of 'follow the road for two days and then turn left' kind of journey – over a cattle grid that rattled your teeth and your suspension and into the yard. There, a large and ominous boar lay in wait where there should have been a welcome mat. I think Ivan took a perverse pleasure in the deterrent effect of the boar upon itinerant worm reps, insurance and double-glazing salesmen who might otherwise have intruded on his routine.

The boar waddled across to the Land Rover and snuffled about the wheels; unlike collie dogs on other farms it did not cock a leg on the hubcaps, just peered at me down its very long snout. I waited in the vehicle until Ivan gave an assurance that the brute had been fed before getting out, 'When in doubt do nowt' being a favourite maxim, in spite of my southern origins, that has kept me in one piece over the past twenty-four very odd years. Once the all clear sounded, I disembarked and saw an elderly sow nursing a litter of piglets snuggled up in the remains of last year's hay and straw in the Dutch barn. I think there were thirteen but it may have been twelve with one running from front teat to rear and getting counted twice. I couldn't say for sure.

The pigs were as free range as it is possible for animals to be; they had the run of the farm, the build-ings and the woodland beyond. In autumn, they would trek across to the woods in the morning and come back at night, sated with acorns. Only two areas were strictly out of bounds: 'They're not allowed on the leek bed and they're not allowed in the house either.' Ivan took his leeks very seriously, winning valuable prizes with them. As for the house, well, the workings of its innards were probably a bit of a mystery to him.

We went through the ritual of examining the pigs, asking Ivan which ones he thought would do us best and then 'choosing' them. This was a complete charade because we never chose, but always accepted his recommendations unquestioningly. The reason for this was simple: Ivan was a good judge of stock. Even more important, Ivan would not have known how to diddle anybody. Congenitally incapable of duplicity, he would never have made much of a worm rep or salesman himself. You meet very few people like Ivan these days.

After several of these trips up to Roadhead, we

changed our Land Rover for a car, and although the back seat of a Ford Escort is very nicely upholstered and the stereo supplied all round sound, I baulked at the idea of putting four little pigs in there. No problem, said Ivan: he would fetch the pigs to Rowfoot for us. Gratefully, I suggested that Ivan might like to stay for his lunch. Yes he would. And he'd arrive at twelve.

Inexplicably, I cooked stuffed hearts for lunch. The Rayburn must still have been belching away, for I am fairly sure that I cooked them in the bottom oven. Stuffed hearts are great so I'll just give you the recipe: take some lamb hearts, snip out the gristly sinews and fill the cavity with decent sausage meat. Stand them on end, pointy bits upwards, have a ferret about in the fridge to see if you have any tasty veggies that need using up – a stringy bit of celery is good, as is a carrot with an incipient bloom of penicillin, or a couple of onions sending green shoots into the shelf above, that sort of thing – chop the whole lot into dice and scatter them around the hearts. Then sling in a goodly slug of dry cider, some stock (proper stuff made from bones, please, none of those nasty tiles quarried from monosodium glutamate), and any herbs you can find. Put the lid on the casserole – did I mention the casserole? – shove it in the bottom oven of the Rayburn and leave it overnight. By morning the kitchen will have filled with a sweet meaty aroma and the dogs will be salivating enthusiastically. Let them out. It is not for them. Remove the casserole from the Rayburn and lift the lid to reveal as flavoursome a feast as you are ever likely to eat.

I know lots of people don't like offal, but if you can't face eating the gubbins, then you are wasting some of the best bits of the animal. Sweetbreads are divine, though I'll pass on the kidneys, if you don't mind. I am sorry, but I can never totally separate kidneys from the job they do. I am always reminded of it when we kill a sheep at home. Somehow, no matter how carefully you hang the carcass up, there is always, always a tiny drop of wee left in the kidney when you pluck it from its fatty berth. So no kidneys for me, thanks. And the EU comes over all sensitive about brains (I was just astonished to discover that my sheep actually have any) and tongues. When not measuring angles on bananas and cucumbers and dealing with shocking human rights abuses relating to school uniform, the bureaucrats in Brussels turn to thinking sinister thoughts about offal.

So we sat down to the stuffed hearts and Ivan, bless him, said I was a 'terrific' cook. I glowed with pride, having had an enormous complex about the subject since being effectively thrown out of cooking classes at school. It gets worse. At our school reunion the teacher who had ejected me so summarily asked, 'Did you ever marry?' 'Oh yes,' I replied, and before I could continue she barked, 'Is he still alive?' That Ivan had been impressed was really most rewarding and I thought about writing to the cookery teacher to alert her to my progress. I think she would be glad to know that I was just a late developer.

The next year, we didn't bother trudging off to Roadhead at all. We just rang Ivan up and ordered four piglets much as most people would fill in their order form for Tesco Online. Ivan said he would bring them. We said that would be great and we'd see him Tuesday. Minutes after the receiver had been replaced in its cradle (this was before the days of walk-about-with-them-or-drop-them-in-the-bath models) the phone trilled again. It was Ivan. 'Can I have me dinner?' he asked. 'And can I have what I had last time, 'cause I fair enjoyed that?'

I assured him that both requests would be accommodated.

Tuesday evening was clear and fair and Ivan rolled up in a bright blue pick-up. And to my horror, Ivan and another fellow got out. There is just one problem when you cook hearts. Three hearts are three hearts. They will not, cannot, be persuaded to look like four portions.

I need not have panicked. Ivan and his sidekick unloaded four squealing, kicking little piglets and put them in the shed and then Ivan got round to the formalities.

'This is Tommy. And Tommy's going to the pub. Bye-bye, Tommy.'

Tommy looked ever so slightly surprised to learn that he was going to the pub and even more astonished when Ivan said he was to be back, without fail or delay, to collect him at half past ten precisely. Ivan had a good night in mind. And so it came to pass: down at the local poor Tommy incurred the wrath of a neighbour who was on a promise with the barmaid there and thought Tommy was trespassing on his territory, but at Rowfoot it was a night filled with discussion, laughter and some pretty slanderous gossip. And the hearts were delicious, by the way.

The system never varied thereafter; Ivan brought the pigs, and we had hearts for dinner.

Our Gloucester Old Spots lived out in the garth and up to their nickname of Orchard Pigs. They rooted about in mud, slept one on top of another in their sty – codged up from the remains of an old piano and a kitchen bench, with a couple of fertiliser bags nailed to the roof for waterproofing – and scoffed the fruit they had dislodged from the trees with a special kind of job satisfaction. And they ate the entire orchard, not just apples but bark, sap and leaves too.

The Tamworth crosses took rather longer to finish than the Gloucesters due to a spell of cold weather, and fate, or rather Nat, intervened.

'We always had pigs at Parkhead,' he said. 'Used to cure our own bacon.'

My ears pricked up like a lurcher's.

'Often thought we should have pigs again.'

'Really?'

'Course, we could do it between us.'

Of course we could. Silly me, why didn't I think of that?

In no time, somehow we had arrived at a Plan. The Tamworths would be shipped down to Ainstable Hall, Nat and Jean's farm just down the hill, so near that had we really wanted to we could have communicated by megaphone rather than telephone. Proximity notwithstanding, we had a fair logistical problem here. It is no good whatever saying to a pig, 'See, down there; just down the hill and take a left into the drive,' not because the pig won't understand – he will, and very well – but because, in normal circumstances, pigs are disinclined to take orders from humans. It just does not work like that: pigs know better. They know that further on, over the bridge and right, say, is a great deal more interesting, so don't ever think of trying to move your pigs without assistance – think along the lines of three good men and true – gates, hurdles, a trailer and considerable patience.

We agreed that we would continue to supply the skimmed milk, ferrying it down by the bucketful each evening for the delectation of the piggies. Nat, meanwhile, was tickled pink as he had located a source of Free Food: his tractor man had a friend whose father (are you still with me?) had one of those hot baked potato stands and often ended up with unsold, unusable potatoes at the end of a hard day out on the hustings. He threw them away. Well, this was obviously a terrible waste – but hold! That situation could be remedied. Potato Man's social conscience was not beyond redemption. He could be saved – by our pigs, natch.

The tractor man would bring the cold and crinkled baked spuds to work with him each day and Nat would deliver these goodies to the pigs, who, having an impressive commitment to the principle of recycling, would be suitably grateful.

And do you know, the Plan worked wonderfully well. On a diet of skim, spuds and a bit of rolled barley, the pigs prospered. Like Topsy, they growed. And growed. And growed some more. They were past being pigs now; they were more like some forgotten prehistoric species that roamed the earth in search of a small but well-fleshed child for breakfast and a wingeing adolescent for lunch. They were monsters.

'When do you think the pigs'll be ready?' I asked Nat.

'We'll give 'em a laall bit longer,' he would reply.

This went on for some time.

Finally, Nat deemed them ready – they were outgrowing their shed and beginning to tinker with some loose brickwork – so the deed was done. Strung up from a lofty beam in a barn at the Hall, their trotters touched the floor: that's how huge they were.

It has always been legal and, I would argue, advisable to kill stock at home if they are for your own consumption. They were indeed for our own consumption – possibly until sometime in the next millennium, I thought as I looked at the size of the carcasses.

Anyone of a delicate disposition, vegetarians and the faintly queasy might do well to skip the next few paragraphs, but hang on in there if you possibly can.

The first thing is to scrape the spiky hair from the skin, chiefly because you do not really want hairy bits in your crackling. It will give you indigestion. If the pigs have lived outside they will have grown a coat rather like that of a soft-hearted porcupine; removal of this cladding is messy and you will need vats of boiling water, knives, cloths and lots of energy.

If you are intending to make black pudding, you will need to bleed the pig. You'll need to bleed it anyway, so why throw the proceeds away? Black pudding is just pigs' blood, some chopped fat, seasoning and a bit of mint torn up and thrown into the mix; we did ours in the Rayburn, finding it congealed agreeably after an hour or so. Slowly slowly catcheee piggeee.

The first snag is, of course, that you have to get the blood from pig to pot. The safest way seemed to be to carry it in a bucket and just walk up the hill homewards, the idea of vehicular transportation being far too dangerous: imagine putting the bucket in the well of the Land Rover and hitting a pothole. The stuff would slosh everywhere, leaving the inside of the vehicle looking as if it had been involved in a violent road traffic accident and quite possibly putting your dogs off travelling in it ever again. Still, you wouldn't be troubled by hitchhikers, would you?

So I set off up the hill, bucket in my left hand. And do you know what? I could hear my mother's voice ringing in my ears: 'I wish you'd get a proper job, Jacqueline.'

Black pudding sorted, the next day Jean and I convened to make sausage while Nat began burying the joints in a mix of salt, brown sugar and saltpetre. The last ingredient had been the hardest to find, since the scientific development department of the IRA had found that saltpetre made their explosive devices detonate with that extra satisfying little snap, crackle and pop. So retail chemists, being of a sensitive disposition (so probably not reading this bit anyway), had become reluctant to sell it over the counter. And I suppose when a red-haired, green-eyed female turned up waving a copy of *Belfast Today* and asking for saltpetre it made them especially nervous, so we sent a dark-haired, bearded, not-Irish-at-all neighbour to source the saltpetre and told him not to wear his black beret or his balaclava for the foray: a trilby would do nicely.

It is worrying how gullible retail chemists are, as he bought plenty without any difficulty. I shall have to send him back for the sulphuric acid I need for curing sheepskins, before I get the bath resurfaced. Sulphuric acid plays havoc with your bath enamel.

Jean had been busy organising the other vital bit of shopping for Project Pig. She had acquired the necessary sausage skins from a local butcher though I have no idea how she explained her requirements or whether she ordered them by the half mile or by weight. With two Kenwood Chefs and their mincing attachments, we worked through pound upon pound of pork, accumulating a mountain of mince on the worktop. Then we broke for tea.

'They used to use proper pigs' guts for sausage skins, all washed and cleaned out. These 'uns is plastic.'

Oh, the shame of it.

Our disgust was as phoney as the plastic – I mean, who wants to wash out miles of guts with brine – so we giggled. As Jean fixed the sausage-filling attachment, an obscenely fashioned plastic tube, on to one of the Kenwoods, Nat dawdled in from outside, hat slightly skewed on his head, baler twine holding up his elderly trousers and, if I wasn't mistaken, sheep shit all over his hands: the archetypal Cumbrian farmer. He had sniffed out the tea and decided to take his break too.

'Do you know what the Victorians used sausage skins for?' I ventured.

'No, what?' Jean clearly had not come across this little nugget of entirely useless information before.

'Contraceptives, that's what.'

Jean looked first at Nat and then at me and I know it's pathetic and I know it's in terrible taste, but we fell about laughing.

Small sweet pigs' cheeks, great hams and sides of bacon eventually emerged from their salty tombs in the back kitchen and were tied up to dry. The hams alone weighed several stones and having refused several kings' ransoms to part with them and being conscious that for the first time in our lives we had something worth stealing we took the only sensible course of action we could.

We drew the kitchen curtains at night to conceal our valuables from prying eyes.

We have snow.

Blackbird and robin have impressed their fragile prints on the crisp snowfall on the cobbled yard. On the hill, crystalline snowflakes that fell through the night have stayed for breakfast; jewelled fragments of ice glitter and sparkle under a brilliant yet chill sun. The ewes scamper across the miniature ice floes that yesterday were limpid puddles around the feed troughs, gratefully pushing their snouts into the deep chewy mass of molassed meal. The one who is the model for the new range of Swish curtain poles uses her horns to clout all comers out of her way and sneak extra rations: a wicked ploy but a devilishly effective one.

The dogs, as usual, ignore the sheep, pee enthusiastically in the fresh snow and leave tiny, perfect and vaguely yellowing circles in their wake: a frozen bucolic February fantasy.

Life could be worse.

March

*March winds snap and bite, pierce and prick as
keenly as any harpoon, but gradually the route of
our Sunday walks alongside the Eden starts to come
alive again; catkins dangle on boughs, daffodils are
in bud and the snowdrops, heads bowed now, are
fading. Thin green needles of grass thread their
way through the tilth, goodness stored over winter
fuelling their inexorable rise. There's a wonderful
sense of promise at this time of the year: a
reawakening, a rebirth.*

MARCH IS A schizophrenic kind of month, not
entirely sure whether it is the end of winter or the
beginning of spring. The weather oscillates between
bursts of extended sunshine, when small lambs dance
dangerously on their mothers' spines, testing maternal
patience and getting a fresh perspective on their
surroundings, and at t'other end of the meteorological
spectrum glacial shards of rain spiking the back of your
neck as you struggle to stay upright under heavy battery
from gales that haven't paused for breath since leaving
places known only to listeners of the Shipping Forecast.

March is synonymous with madness: symptoms include sitting outside on Sunday mornings tut-tutting over the newspapers and swopping Ugg boots for open-toed sandals. That's on the good days, obviously. And on the bad ones, at least there is confidence that however bad it gets, it won't be for long.

As a kid, I was packed off to Suffolk for the Easter holidays. My enigmatically named Auntie Grey met me at Stowmarket station and always boomed exactly the same greeting: '"Spring in the air, Colonel!" "Dammit, why should I?"' before bursting into peals of laughter that echoed round the station, sending small birds skywards and ticket collectors scurrying in search of earplugs. Spring, she reckoned, was all that was wonderful, but she was a woman in the permanent grip of such uncommon cheeriness that she reckoned most things were wonderful. Auntie Grey was a little bit potty; she had a 'glass place' not a conservatory and kept a long, murderous sword by the bed. As far as I know she never used it but I cannot be entirely sure.

I passed those train journeys between Liverpool Street and Stowmarket in an absorbing little game: from a base of nil, I added points for fields with ponies in them and deducted points for fields with cows. No idea what I did about sheep or free-range pig units, much less garden centres and trail bike tracks. Then again, garden centres used to be called nurseries in the sixties and trail bikes hadn't been dreamed up at all.

The clichés of spring – gambolling lambs and Wordsworthian daffodils – affect me far less than the small things that I came to appreciate during those childhood holidays. Returning swallows, the smell of freshly mown grass, and the first primroses the colour of melted Jersey butter growing in profusion alongside the paths of Coombs woods by the Eden, swirling cheerfully up gradual inclines and tumbling down becksides, transport me straight back to Suffolk and the woods around, coincidentally, Coombs Church.

Back in London after Easter, there was usually enough grass for the riding school ponies to be turned out in the Priory fields – yes, *that* Priory, where nowadays fallen stars nurse bruised egos and combat interesting addictions. However idiosyncratic its architecture – and The Priory is a fine example of Strawberry Hill Gothic Victorian with some wild crenellation – the afflictions of its inmates in the noughties are even more fabulous: persistent text offending, for example, is the medical term for people who text other people for up to seven hours a day.

We didn't text, as mobiles, like trail bikes, hadn't been invented; we just flung headcollars on the ponies, jumped on their backs and rode across a busy B road to the Priory fields with scant regard for our personal safety – I cannot even recall that we wore hard hats – and hurtled a couple of circuits. Usually someone fell off but we bounced in those days and even broken arms had welcome implications for excuses in relation to maths homework for up to six weeks, although none at all for avoiding prosaic household chores like removing the tidemarks round the bath. And we never even considered the threat of long-term arthritis. Still, age has its compensations: there's comforting draughts of Wincarnis and discounts on rail travel to look forward to, plus the heady prospect of selecting a stair lift from the Sunday colour supplements.

Spring is turnout time and wintered-in youngstock are released into fields for the first time. Agoraphobic calves are not unknown: I had a little Limmy bullock who liked his old sandstone shed (I don't blame him; it was much prettier than the hideous hangars beloved of modern agribusiness) so much that he never wanted to leave. He probably thought he had been on some Care in the Community programme for the past five months with regular meals and twice daily room service and had become immovably attached to the place. I tried chasing him, waving my arms in the air, pulling on a pair of Wallace and Gromit atomic trousers and yelling at him: nothing shifted him. In the end, I shoved my shoulder under his backside and heaved (don't try this unless you are wearing a very unloved sweater) and he sat on me, and eventually we struggled out of the byre and into the spring sunshine together. We made very slow progress for a while, and then, quite inexplicably, he eschewed the wide inviting open gateway into the field containing his mates and leapt a solid five-bar job to the right of the gap instead. Then he planted himself for some time, quivering and surveying his new surroundings of half dead spinach plants and artfully rotovated soil. That was 1997 but he was so agoraphobic that he might still be there: I'd better go and look.

Limmy bullocks, you may infer from this tale, have poorly developed directional skills so it is no use pulling over on to the hard shoulder and asking one grazing in a field adjacent to the M6 how far it is to Shap. But if you ever come across a bullock with similar steeple-chasing ability, buy it, take it home and put a saddle on it first and your shirt on it when it goes down to the start at Kelso: you know it makes sense. And just one more small point: they are Limousin cattle, from the Limousin region of France – roughly bisected by the sauntering, watery Dordogne – not limousines which are very fancy cars that the people who breed Limousins tootle about in. The cars do not differ from the cattle much in cost but they have better traction.

Spring is unique amongst the seasons. It can't be preserved like summer jam, pickled like onions in autumn or bottled like damson gin, as surely the spirit of Cumbrian Christmas as goodwill itself. Spring is filled with such expectation of plenty that no insurance against want seems necessary.

And now, spring is just around the corner. That's left at Rowfoot, past the chapel and down by the litter bin to be absolutely accurate.

During this calm-before-storm time, it seems a good idea to decorate the front bedroom. The trouble is that if we leave it until the cottage-letting season gets going it begins to feel as if we will either never start because we are too busy or never stop if we do ever get started . . . so off we go to buy wallpaper, paint, those little things that you peg back phone wire with, paste and all the gubbins. We go to B&Q. I only let Malcolm go shopping in there on Wednesdays because that is geriatric discount day and he gets 10 per cent off: one of the multifarious benefits of marrying an older man is that you save on your home decorating costs.

It all starts well enough. I play about with the wallpaper stripping machine, occasionally singeing my wrists with boiling water but not inflicting any really serious damage. The paper comes off the walls in satisfying strips. It is all a long way from when we first moved to Rowfoot and thought dangerous thoughts of getting straight for Christmas, in our haste doing every room with woodchip. Well, it was so cheap. Why was it cheap? Because once on, it becomes welded to the walls and is nigh on impossible to shift, that's why. You only discover this when you come to redecorate and then you spend fully four days scraping the wretched stuff off in three layers – first the paint, then the chips, which insinuate themselves under your fingernails like splinters, and finally the backing parchment. You will vow never to do this again. Ever.

When we came to do the hall – it's not a hall, really, just a space at the bottom of the stairs – I refused to do it manually, instead lashing out the absurdly extravagant sum of eight quid on the hire of a steam machine for a weekend. Cheaper than divorce, I said at the time.

Malcolm is, I think, allergic to decorating. He's contracted sciatica again and is hopping about on crutches with a very pained expression. The bedroom could take some time.

A fortnight later: it is not finished.

*　　*　　*

A week further on and it is not finished now either. Never mind. I have completed a distance learning course (Malcolm shouting from the bedroom) in how to use a Bosch drill. The sheer exhilaration of finding that my screw went into the Rawlplug, into the hole I had drilled all by myself, was quite unlike any emotion I had experienced before. I could go for this DIY stuff. I could get good at it. I could be a joiner, like Eric. This, on balance, is unlikely since Eric served his apprenticeship making coffins and that is why, he says, he always measures twice and cuts once. I had better content myself with fitting Rawlplugs and not get too over-excited by my early success, hadn't I?

This time in 2001, foot and mouth ripped through Eden, this tranquil land of river valleys, sandstone gorges and rich stockrearing pastures, tearing the heart out of its bucolic charm. One point three million head of livestock were slaughtered and acrid smoke from pyres engulfed the area; four years and millions of pounds on, the view across the Eden valley today, as sheep and cattle graze peacefully once again, suggests that the area has recovered its equilibrium, that everything is back to normal.

The reality is more complex.

Anxious to get back to what they knew best, most farmers passed up the opportunity for a fundamental re-evaluation in the enforced breathing space of the pandemic's aftermath. Some, though, have diversified: ice-cream parlours and farm shops, golf courses for people and cross country ones for horses, swimming pools and gyms popped up across Cumbria, funded by DEFRA's generous recovery packages – or the government's thirty pieces of silver, depending on your cynicism threshold.

Others, relieved of making a conscious decision about their future and cushioned by compensation, retired or sold up. Some were just too disillusioned to carry on: certainly those who fed beef animals only to find that when movement restrictions were lifted they were over thirty months old and worthless had little incentive to stay within the industry.

The majority made a swift return to farming, abandoning resolutions to buy co-operatively in the rush to restock. Movement restrictions meant that many dairy cattle that should really have been culled – kickers, cows prone to mastitis or lameness – were put in calf again. Predictably, neither they nor their offspring have been valued specimens. Elsewhere, opportunist breeders tried to get their slice of the compensation cake by inflating prices to absurd levels, and as one farmer commented pithily: 'Daft buggers like us were in such a rush to buy that we paid them.' And then there were some that just 'shrank when we got 'em home . . .' It has not been a straightforward process.

Since then, milk prices have plummeted; to maintain the same income, farmers need to milk more cows. Or, to put it another way, the fun has gone out of the job. 'Farming is not a way of life any more, just a way of making a living,' says one dairyman. 'I always used to have a couple of cows I had a real soft spot for but they went with foot and mouth. These don't feel like mine.

They're just cattle I've bought, not cattle I've bred, whose mothers I've milked.'

Progressive farmers have been able to regenerate the herds they lost by the use of pioneering embryo transfer technology but this practice is neither widespread nor cheap; now that its advantages are clear it will be interesting to see if it becomes more popular. The Rare Breeds Survival Trust has recognised the need to insure against catastrophic loss of important bloodlines and has set up a gene bank, but ironically the government's National Scrapie Plan looks likely to decimate more bloodlines than FMD!

Rules and regulations have increased since FMD. And even though I failed to find anyone who actually liked increased paperwork, it is generally agreed that management has improved. 'We wash out our trailers a darned sight more often,' said one well-known sheep breeder, while my dairying neighbour said: 'We are all much more watchful . . . and we have less muck about the place, or on our wellies!' Undoubtedly record keeping has improved, pedigrees are better documented, and animals are more easily traceable – vital in times of increased consumer sensitivity.

FMD's swift spread also threw into sharp focus quite how often, how far and how carelessly we were moving animals about the country, and, while the new transport and six-day standstill rules go some way to closing those chinks in the armour, taken together with the thirty-month rule they mean that forward planning of stock movements is a new imperative. Auction mart closures elsewhere have done nothing to diminish enthusiastic trading through Cumbrian sales rings, leaving the county particularly vulnerable to future outbreaks of livestock diseases. And while biosecurity is a priority at auctions nowadays, more stringent controls at ports, such as apply in Australia and New Zealand, inexplicably have not been implemented here.

One completely unexpected consequence of the pandemic has been the erection of miles of fencing in the Lake District uplands, vital to re-establish the hefted flocks of Herdwick and Rough Fell sheep. How these flocks fare remains to be evaluated, as Rough Fells do not appear to thrive away from their natural heft (the fell on which they were born and raised, the place where they will do best). Like the North Ronaldsay, they rely on their native environment for dietary balance.

The bigger picture is worrying. Vets concur that the disease profile of Cumbria has altered, but not for the better. As Cumbrian farmers sourced unprecedented numbers of cattle from across the UK, so the incidence of TB increased. Before 2001 this was not a significant local problem but in 2004 alone eighteen farm premises were affected, including one farm culled out after TB testing that had previously been culled with FMD; only time will tell whether farmers are becoming inured to repetitious misfortune or whether they are broken by it.

And questions – questions that needed answers in 2001 – still remain unresolved today.

How did it start? In the absence of a public inquiry or in-depth investigation, we are none the wiser. Ethnic restaurants, illegally imported chimpanzee meat, discarded sandwiches from flights into Newcastle were all blamed briefly but no definitive answer has ever emerged. We continue, though, to import meat from countries where FMD is endemic – indeed, we feed our army on it – and thankfully without recurrence. But we still don't know how the disease arrived.

How does it spread? Does the virus travel by air and if so did the pyres spread the very thing they were designed to eradicate? Do birds carry it? Can it be carried by people, on their boots or their clothes? Did the lorries ferrying carcasses spread the disease? Is it transmitted animal to animal? Which, if any, disinfectants kill the virus and were those mats and sprays really any use at all – or were they used more in hope than in expectation, like so many of our panic-fuelled actions and reactions at the time?

What about vaccination? We're still thinking about that, apparently . . .

We know now precisely what we have always known: FMD spreads with quite terrifying rapidity.

Most shockingly, there is still no National Contingency Plan. The National Farmers' Union have formulated a plan for Cumbria in consultation with local vets, Emergency Planning Officers, councils and DEFRA, who have been laudably co-operative. Regrettably, though, until a National Contingency Plan is agreed with the State Veterinary Service and fully in place, practical and logistical reasons would prevent any Cumbrian Plan's being implemented. London, the NFU assure me, is catching up. It needs to, fast.

We need a short, simple plan to be communicated to everyone with livestock. Not just members of societies, organisations and clubs – everyone. Something to kick in instantly: an immediate stop on all livestock movements, sales and auctions would be a start. But none of this is of the slightest use unless we unravel the mysteries of how FMD started and how it is spread. Until those issues are resolved, the disease remains something akin to a viral terrorist, a secret enemy that we fail to understand. That failure leaves the UK exposed and vulnerable.

Before any politician dares chant that tedious mantra 'Lessons have been learnt' it is worth remembering that we have been here before. Lessons were learnt from the sixties outbreak; the trouble is that the reports containing them were neatly filed away and none of the recommendations implemented.

That mistake must not be repeated.

Let us just hope it is not yet too late.

There is always, hidden amidst the misery and the anguish, a funny side. It might not seem so at the time but it's in there somewhere. So, as we approach the dubious 'anniversary' of the benighted spring of 2001, here, in no particular order, are some of the dafter episodes. They do make me smile, occasionally wryly, mostly knowingly, always incredulously. See if they ring your bell too.

The pandemic is at its zenith. A man is having what we will call a modest domestic bonfire. It is harmless enough – probably no bigger, no more exciting, than the ones my granny used to have (admittedly to the consternation of some of our neighbours in suburban

London who exhibited quite irrational fears for their garden fences) once a year. Being a responsible soul, Bonfire Man has lit his fire carefully and contained it sensibly on a day with no significant wind. It is puffing away as satisfactorily as a Cuban cigar when an inspector calls and says: 'That fire, sir, is causing a nuisance under Section 46C of the Naughty Bonfire Regulations and I demand it be extinguished.' Or something very like it. From his vantage point our hero can see one, two, three, four, five pyres spewing out acrid smoke and points out that, although he is not a religious man, the panorama invites comparison with the high-rise, full-strength conflagrations mentioned in the Second Book of Chronicles, chapter 7 verse v: 'And King Solomon offered a sacrifice of twenty and two thousand oxen and a hundred and twenty thousand sheep . . .' Our man gestures incredulously. Blessed with all the incisiveness of the speaking clock, the official shakes his head and says: 'Different department, them.'

But don't worry. The surgery required to reinstate Bonfire Man's jaw to its rightful place was performed under local anaesthetic.

Just a mile or so distant from the bonfire is a man doing his best to deal with a different sort of wood – sleepers. Specifically, the ones covering the floor of the cattle shed. They'll have to come out. Be burnt. Incinerated and replaced – all in the interests of biosecurity. No, wait a minute, says the man in the white suit, we could take the end ones up, post some thin men down the holes, get them to power-spray and disinfect from underneath, and then replace the ends – carefully cleaned, of course. Or perhaps we should take them all out, disinfect them, and reinstate them afterwards. Would that be better? Oh, hang it, it is all going to cost a fortune. Let's just spray them where they are and hope for the best . . . Or, even better, let's not bother at all – let's forget them altogether. After all, the infected cattle were half a mile away at the time and had never been in the shed or on the blinkin' slats.

Three months pass. Sunshine follows showers, people come and go, everything except the dogs is steam-cleaned, disinfected or resurfaced. And another member of the MAFFia arrives. 'What's these here spots?' he asks, kneeling down and examining specks which might be cement drips, bird droppings or lichen with a peculiar predilection for living in an extravagantly disinfected environment. He decrees that every speck must be circled with white chalk. And scrubbed with a wire brush.

And so it came to pass . . . but no-one ever knew why.

One normal thing occurred during summer 2001: haytime. The stock have gone, but the grass still grows and now it needs cutting, baling, leading – you know, all the things you do with hay. But the owner of this field of hay is not a regular farmer, just an enthusiastic amateur who has to fit the feeding, shepherding, haymaking, and all that stuff in between a full-time job. 'In between' means after five o'clock, generally. But the officials whose presence is mandatory to police all disinfecting, spraying and squirting of the haytime machinery do not work 'in between' hours. They work nine to five. So what happens?

'Well,' they say with great solemnity, 'we'll have to trust you on that one if it's in the evening.'

Being diligent peasants, of course, they did disinfect, spray and squirt, but if they hadn't, no-one would ever have known.

As surely as night follows day, harvest follows hay. Another farmer – not the same one just trying to be clever – needs to combine his barley. It is probably the only saleable commodity he will be able to lay his hands on this year – so he rings up MAFF and says, 'I want to combine my barley. Is that OK with you?' MAFF pause, suck their teeth a lot, scratch their heads and say: 'Ish. But first you must have a licence in order to take the combine harvester into the field.' That's all right then, thinks the farmer, and he fills out the forms in triplicate, as you do, and sends them off. MAFF send him his licence and all is well in his world. Drenched in summer sun, the balmy air is honeyed and humming with bees, and a heat haze rises up in soft billows. The combine whines its way through the barley like an infant in a supermarket. Briefly, our farmer is a happy man. Until MAFF tell him that actually, hang on there son, you can't just take the combine *out* of the field, you know. You need a licence for that. Another, different licence, different forms, you see . . .

We are fed up. We think it might be nice to go out for a drive with Sam. It's a sunny day. But first, perhaps it would be an idea to ring up MAFF and just check that we're not going to contravene a regulation that someone has thought of while we've been asleep.

I ring the number. 'I'd like to take my horse out for a drive. Any special requirements – a licence, perhaps, or should I push him headfirst into a pool of disinfectant before setting off?'

'A drive, madam? Well, if you pick his feet out and then take him out in the trailer and sort of drive round a bit and then come back and unload him at the premises you started out from – I can't see any problems there at all.' I explain, patiently, that we are not Wallace and flaming Gromit, having a 'nice run oot', we are going for a *drive*. A chariot, a cart, a trap, a gig is involved here, not a blinking trailer.

Before the telephone wire spontaneously combusts I am speaking to a South African vet. Nice bloke. Says he's sorry but he hasn't the slightest clue, but if Sam's feet are scrubbed clean and no-one actually prostrates themselves in front of the beast, he supposes the best thing I can do is just go right ahead. And see what happens.

I survived to tell the tale. This could be because no-one else knew what the regulations were either or it could be that they were so glad to see something four-legged and alive that they didn't care. I'll plump for the latter explanation, myself.

A year on, and you might be forgiven for thinking that slavish dedication to rules was becoming slightly passé. Not a bit of it. The restocks have started to lamb and it is a cold, wet Sunday. Sunday being a weekend, previous experience suggests that it is a trifle unlikely that we will be able to locate an official to sanction a movement order. So there's a bit of a moral dilemma: do you leave the lamb to die in the absence of a movement order or do you bring it inside? If lamb and shed were both on your own holding there would be no problem, but in accordance with the principles of Sod's Law the lamb is on a bit of rented turf: outlying land not your own. And anyway, there is a slight difficulty with the sheds on your own holding since MAFF told you to take them down and burn them and have not quite made up their mind whether or where you can build new ones . . .

The final joke is on the British taxpayer. The European Commission has told our Chancellor of the Exchequer Prudence Brown that it is withholding £600 million of the money claimed in 'expenses' by the Treasury from Brussels. Why? Because the so-called pre-emptive cull was illegal as MAFF had no authority to slaughter animals unless directly exposed to infection. That's about £25 per taxpayer, shared out. And here in Cumbria, in 2004 DEFRA (*née* MAFF) still owed about £7 million to firms involved in the FMD clear-up and clean-up operations.

Before I succumb to terminal sense of humour failure I cheer myself up by recalling the ministry official who asked a Monmouthshire farmer's wife: 'Is your bull male or female?'

You couldn't make it up, could you?

It's that time of year again: time for the lambing forecast.

It's the bleakest day for weeks. It's pelting, there's an angry gale blowing and I have a cold. 'Been sleeping with your feet out the window again?' enquires Freddie.

'Only on Mondays,' I splutter. He is cladding himself in waterproofs, a hat with furry earphones and his wellies and, for reasons he keeps to himself, he is doing this in the coalshed.

The sheep are already captured. They chose to be particularly co-operative this morning and went where they should when they should. I have come to the conclusion that it is best to treat sheep like men: suggest, persuade, cajole if you absolutely have to and then wait in hope. With luck, they will decide to do exactly what you want them to, only they will think they made an independent, enlightened decision all on their own. It has taken me twenty-four years to work this out. I may not be the sharpest tool in the box.

So I fed them, left the gate open and waited. Experience has taught them that when A N Other, or indeed Freddie, appears, the resultant experience is likely to involve medicine, injections, and/or subjection to physical indignities like having your toenails cut, and all of it is as unpleasant as it is avoidable. So, learning from precedence, they steer well clear of any human but me and I am only tolerable when I have a bucket in one hand and a crook in the other. But today the only wobble occurred when Curly Horns momentarily questioned the wisdom of getting her hooves wet in the muddied gateway. The only predictable thing about Curly Horns is that she's always devilishly unpredictable and very stroppy. She has a low panic threshold and would argue with a signpost. We are not natural allies, she and me.

Anyway, here they are.

The wind is bending the branches on the garden lilac into oxbows; the laburnum looks more lachrymose than any willow, as its fragile tendrils flutter like tiny pennants in the gusts. The fierce wind casts bird feeders from their berths, tossing them on to the lawn where they shudder in palsied shock.

We dose and inject the tups first, then pen them behind tied hurdles. The ewes take rather longer, grownups first. They all appear to be safely in lamb and, so far, nothing that shouldn't be isn't. That's good. It's tricky injecting them at this stage of the year because their fleeces are so very dense. It might be imagination but they seem thicker than ever this year and it has been the hardest winter for some years. Coincidence? Or perhaps they knew something we didn't. It wouldn't be the first time.

'It'd be a bugger if the end came off the syringe,' I muse.

'Never find it in that lot,' agrees Freddie.

Syringes and ends remain connected today, thankfully.

We dose them with wormer, stuff for ordinary run-of-the-mill worms and fluke too, plus selenium and cobalt, deficient in our own land – so it's no-expense-spared stuff, this. It has been specially prepared partly to compensate for our specific environmental shortcomings and partly because all the companies that make sheep wormer are monoliths themselves and seem unaware that all sheep keepers aren't the same. It has not occurred to any of them, seemingly, that anyone shepherds fewer than several hundred sheep requiring several hundred doses of wormer dispensed from an enormous plastic satchel. So good old Carrs, our local agricultural supplies people, make me up some bespoke stuff. And although it might look like a strawberry smoothie, it sure tastes a whole lot different as I find out to my cost when I hold one syringe in my teeth while I shove some multi-vit down a gullet with another. Vile.

When we finish we actually have some doses left. I always order slightly more than I need because I tend to be a bit generous with the measures – I'm the same with whisky but I'm a great deal more careful with the peaty output of far-flung Scottish islands than I am with sheep wormer and I spill one dose in the mud. I tip the residue back into the bottle but the wind whips some away in pink snowdroplets.

By the end of the session Freddie and I are skating about perilously on a slippery surface: we decide not to try to do anything about sheep feet today because we are both soaked to the skin already. Finding their feet in the mire would be difficult and messy and that's before we even start parting their toes, cleaning their hooves, and trimming them. If we were both accomplished mud wrestlers it might be a possibility, but we are not.

One or two of the ewes need their horns trimmed so that they don't curl right round and grow into their faces or, worse, their eyes. This is a Man's Job and I hand the saw to Freddie. My job is just to sit on them to make sure they don't move; funnily enough they stay relatively still throughout.

We have a cursory run through them for eating potential. The tups, with good horn placements, faces and points the colour of Bakelite and tight dark fleeces, are something of an embarrassment of riches this year, the girls the exact opposite. So I think I'll keep The Boy – all antlers and muttonchop sideburns – to do his stuff again this year, castrate (drunk with power, that's me) any boys born this year and run them on for next year's freezer, and, unusually, send three of the gimmers for butchering this back end. The other sort of mutton chops, they'll be, but it is important to be selective and breed from only the best.

Glad to have finished, we decamp, Freddie to the coalshed once again and me to put the kettle on. Then we put the world to rights. That's the easy bit.

Burglars? Chop off their hands. Rapists? Chop off their willies. Murderers? Shoot 'em.

Next?

I have had better days.

It was, though, undeniably a day of high and sustained excitement from start to finish.

The morning started quietly enough: dawn, rime, sunshine, all that. I shouted the tups – Arfur and his sons – down for their breakfast and three sauntered down, leaving one curled up and leering through the fence at the ewes. The unholy trinity tucked into their molassed muesli and number four, with his back to me, just stayed there ogling the foreleg-candy (sheep, you may have noticed, don't do arms). The wind was in the wrong direction, so I had to go up and call him – 'Food, food, food.' My voice was almost carried away on the breeze but he heard enough to be startled into the prospect of missing his rations and immediately leapt to his cloven hooves and scampered down the field leaving the girls bereft.

So that proves it: food beats sex every time.

No doubt countless very expensive research projects would be needed to arrive at this conclusion and they would fill much broadcasting time on Radio 4, but I have reached it at no expense whatsoever to the British taxpayer. The thought gave me a nice warm glow. Though that might have been the porridge.

Later that day, just as my nerve ends were knitting up, I fed the tups. Trouble was I had left them in the pen rather longer than they would have preferred, mainly because I was investigating muffled thuds in the roof and then dispensing blue pellets of mouse poison in porcelain dishes to attract discerning rodents. One of the tups expressed his displeasure at my tardiness by launching a rearguard assault, dealing a devastating blow to my rear which wasn't guarded at all.

I was on the horns of a dilemma: the dilemma was called Arfur, Arfur Chance to give him his full title. And I didn't have one.

I can manage with less excitement than this at my time of life.

Gerald the pet pheasant still visits most days. He plops into the garden at the back of the house each morning and grubs about most rewardingly under the bird feeders hanging from the cherry tree and strung up along the railings. Gerald flies lower than most pheasants since his ingestion of vast quantities of cereal has resulted in a girth rather less aerodynamic than most game birds'. It has been swelled by generous offerings from neighbours, amongst whom is one retired farmer. Old farmers never retire, of course, they just become foremen from a distance and retain a lifetime right of access to all sheds, fields, plantations and stores on their former holding. This is a bonus for Gerald. So charmed has our neighbour become with him that he has taken to pinching food out of the pheasant feeders in the wood to swell his girth still further. Up until quite recently the same farmer's wife had been pilfering from the feed bags on the farm to feed Gerald's barley habit, abandoning the practice only because the barley took root in the lawn and blossomed into great long, hairy grasses.

Gerald stays under our juniper bush on shooting days. He's no bird brain.

There are times when it feels as if your life is not really your life but someone else's and you are an impostor. Worse, you might be rumbled. Today was one of those days.

The *Daily Mail* is serialising my first book, *The Funny Farm*, so something called a photo-shoot has been arranged. The Picture Editor – a lady called Kate who deserves those capital letters – phoned to set this up, adding that they would 'send hair and make-up'. What I really wanted to say was 'Fortunately, I've got my own hair and nearly all my own teeth . . .' What I actually said was 'Oh.'

On the morning of the shoot, to which neither

Gerald nor any other pheasant was invited, Hair and Make-Up (otherwise known as Anthea) phoned from the train saying that the train – and I don't want you to be shocked by this – was running a tad late. 'How will you get from Carlisle to here?' I asked, about to offer to go and pick her up.

'The photographer's meeting me at the station,' she assured me. The *Daily Mail* think of everything. I began to tremble: photographers, stylists . . . it's all a bit much for a one-trick pony in the sticks. Did Anthea *do* ponies?

I would soon know.

Fortunately, Anthea and Alan, stylist and photographer to the stars, turned out to be great fun, even glad to be away from the Beautiful People they usually worked with. There was a momentary wobble when they drew up in the car and I thought they might take one look at the raw material they had to work with for the day and just disappear in a cloud of exhaust fumes and screeching tyres, but they were made of sterner stuff than that.

It took Anthea about an hour and a half to achieve that natural look. My tearaway hair was tamed, rolled, bobbed, blown, dressed with stuff made by Bumble and Bumble in the good ole US of A, picked, flicked and preened. And no, she didn't find any nits. Alan sat patiently throughout, and we chatted over the roar of an industrial hair dryer – the stars, apparently, do not shine as brightly as they do by accident, either. Casual hair is blow-dried, make-up dibbed and dobbed, and everyone, these days, anyway, is Photoshopped. If you are Photoshopped, you instantly lose two stone, and are cleansed of all blemishes, lines and wrinkles. Of course, you look nothing like your real self and that's a bonus too: you can still go shopping in your local high street and not be recognised.

I do not think that I need fret about this.

Alan, used to working with people who know how to stand, pose and walk, had to make do with my shambling, the ponies' idiosyncrasies and Gyp's occasional lapses into bottom-licking mode, but it was all great fun. We did shots leaning on the gate, shots of the ponies in their stables, shots of the sheep coming over the hill like something straight from the set of *The Sound of Music*, sheep eating at the trough, sheep in the pens, dogs in the yard, dogs on the wall, and then formal clothes for what Anthea hilariously called 'hero' shots. The trouble is that everything I possess is dark. 'Can I have a look in your wardrobe?' she asked. Best not, really, Anthea. But she was insistent and we ended up changing rooms first, then clothes, and, as a last resort, the furniture. Finally, we lit half a packet of firelighters in a no-stone-unturned kind of way, to achieve a softly glowing fire in the background.

They left about 4 p.m. Then Fred from the *Cumberland and Westmorland Herald* turned up with his camera. Fred was done in five minutes: stand there with the dog, hold this crook, keep still, that's it, cheerio . . .

They do things differently round here.

I received a postcard from a friend today. Nothing remarkable in that, you might think. But this is written in brown ink and informs me, very seriously, that this week is Compost Awareness Week. Do you have any friends who use brown ink?

Malcolm answers the phone during the afternoons, when I am supposed to be working (or, as he often alleges, playing on t'internet). He has devised a way of fielding the intrusions on his peace from call centres worldwide, trying to persuade him that he has won a holiday and/or needs a kitchen, a change of phone or power supplier, pet insurance, a walk-in bath or funeral plans. He asks them what the weather is like where they are. Then he tells them what it is like here. He asks them whether they are keeping up with the cricket, whether they've been to Mumbai, whether feng shui is sustainable as a lifestyle choice and whether they prefer CNN or BBC news. Then he says that it has been very nice talking to them and wishes them good day. I have warned him he will be blacklisted by a whole range of companies and all he does is grin conspiratorially.

I think he is playing with fire, myself.

If spring means lambing to some people, it means moving house to others. Some just move. Others experience a sudden change of heart about the building they found so appealing when touring it in the company of Twister and Scrounge Estate Agents to the Gentry, and once installed they start tearing it to bits and remodelling the entire edifice, expunging its innate character and replacing it with a homogenised As Seen On TV version.

Either way, at this time of year I keep a beady eye on

the planning applications listed in our local rags. Why? Well, at a time when milk prices have collapsed, sheep retain only a tenuous grasp of their fiscal significance, pigs are as volatile financially as they are aeronautically and the last harvest was so soggy that only anyone with intimate knowledge of flood plain cultivation grew anything recognisable, farming bricks and mortar looks an attractive career move: cement, you see, is so much less demanding than livestock. It needs less feeding yet is far more financially rewarding.

The thing is, land has gone out of production: they are just not making it any more. Rather the reverse, in fact, since bits of this sceptred isle are actually falling off at the extremities owing to either global warming or another of Neptune's hissy fits. Away from the edge, land which is not already under concrete is definitely under threat, as our leaders in Westminster have decreed that we shouldn't be cluttering it up with swine and cowherds, sheep, trees and that funny green stuff that looks like Astroturf with hair extensions. Oh, no. In a drive for creeping urbanisation of the landscape, we should be building houses on it – four million of them to be precise.

And now we really have something to look forward to: PPG (Planning Policy Guidance) note 7 says that new building on virgin sites will be permitted, provided it is 'clearly of the highest quality and is truly outstanding in terms of architecture and landscape design'. The only small difficulty here, of course, is the deciding what is truly outstanding. Wattle, daub, natural stone and hand-hewn slate have been supplanted by

Lego-inspired bungalows with concrete tile roofs and neo-Gothic revivalist garages with roll-over metal doors sufficient to garage a fleet of stretch-limo hearses.

With our livestock industry decimated by plague and pestilence, I know it is tempting to plunder the finite resources, but I still think it is a bad idea. NIMBY? Yes, and not in yours either, mate, come to that. I'm suspicious of anything 'innovative' popping up in the countryside since that worrisome epithet is usually a euphemism for 'hideous and out of kilter with a rural environment' but I am no reliable arbiter of taste. So who is, and can a wild buildfest on this scale ever be reconciled to the rural environment?

Recent deluges have done nothing to deflect probably the daftest proposal of all: siting new housing on – heaven help us – flood plains. The only thing anyone should be allowed to construct on a flood plain is a full-scale replica of Noah's original design. It is not just about what we build, it is whether we should build at all. The countryside is already ravaged by the combined effects of litter, uprooted hedgerows, and loony-bin escapee flail cutters.

Whilst on the subject of litter, will you just permit me a mild diversionary rant? Thank you. Contrary to popular belief, the concept of 'environment' does not start with rhinos and end with whales, but exists just as vitally closer to home – something as low-key and seemingly inconsequential as a litter-pick. Not glamorous, I grant, and certainly not of sufficient international significance for Greenpeace to deploy *Rainbow Warrior* to join in the fun (and since recent floods, even locations that are normally landlocked could have played host to Greenpeace's flagship). On a recent litter-pick, in just under a mile, I collected an entire municipal black plastic bag full of detritus, the greater part of my haul being baler twine and silage wrapping. Such shameful litterbugging is not very Guardians of the Land, is it?

We have built in rural areas since time immemorial: of course we have, and most of it has been seamless enough. It has escaped the notice of most historical commentators, but Hadrian himself had a bit of a run-in with the Development Control boys of the day. Edict XII dated the Ides of March (always ominous) in the year 122 stated: 'With reference to permission granted for a boundary fence of timber from sustainable forests, to delineate the domestic curtilage of Camboglanna Fort, it has come to the notice of this Department that you have in fact constructed a rather large and excessively long solid wall from the Tyne to Bowness-on-Solway, where apparently you ran out of stone. Your imaginative use of indigenous materials and the creation of significant local employment notwithstanding, this wall is considered to be in breach of the conditions appended to your original application . . .'

The noble tradition of flouting regulations not only lives on, it reached art-form status in the west country a few years ago. An Agricultural Worker's Dwelling – traditionally a modest structure in open countryside where housing would not normally have been per-mitted – was to have been built to accommodate a farm worker, vital, you see, to run the farm efficiently. Not that there were any stock involved; it was a fruit farm, but in view of the need to talk to the trees round the clock, it was granted. A few modifications and little tweaks – a portico, a gravelled drive, and an ornamental lake to name a few – were made to the original application and the 'farm worker's dwelling' morphed into a palatial 5,000 square foot Queen Anne style mansion. Chunks of land were sold off, reducing the holding to under twenty acres, and then – you'll never believe this – the 'farmer' applied to have the Agricultural Worker's occupation restriction lifted, his justification being that a holding of just twenty or so acres doesn't need a full-time resident worker.

I am a great admirer of historic agricultural architecture; our own barns here at Rowfoot are rather more attractive than our House that Jack Built (not jerry built, at least) which has been added to, modified and modernised as time has gone on. Conversely, modern agri-structures, those green tin abominations of several thousand square feet apiece, are just livestock factories or repositories for big bales of silage that shed their plastic skins on to local highways.

So, pity the planner who has to oversee applications for any of these unlovely erections. On a good day, he might run into some very nice people with truly appalling taste, or he might be on the receiving end of some extremely creative elastication of the truth – applicants notoriously suffer lapses of memory about whether or not they own land, the potential number of jobs created by the proposed development and the buoyancy (or lack of it) of their existing business, not to mention their own state of health which oscillates between frisky and parlous depending on the requirements of the moment. On a bad day – and it has happened – the poor bloke might even be shot.

Briefly, planners thought it would be a bright idea to devolve their decision making to the local community. They'd listen to local opinion. Ah, but *which* local opinion? Therein lies the rub: do they run with the faction that wants something built, whether they have been persuaded, leant upon or bullied into compliance, or the faction that doesn't? That, my friend, is why committees in distant towns are a very good idea indeed; their very existence obviates the sort of hate campaign that would make men in black suits in Sicily puff out their chests with pride. There is, surely, a sustainable argument for planning applications in, say, Armathwaite, being decided in Exeter: then, rules might be heeded and less blood spilt.

But given the current situation, is it any wonder that planning officers are suspicious?

April

*Ah, the Season of the Fool. My Manxes start
lambing on the first of the month. Could these two
statements be related in any way, do you think?
Occasional gusts and storms send lambs running
for cover, shuddering in the shelter of stone walls,
dyke-backs and hedgerows, but spring is properly
here. Lambs everywhere, in the fields, on the hills,
on the wrong side of fences, in the middle of roads
. . . It's remarkable that they find their way out of
fields so very easily but are congenitally incapable of
finding their way back. I didn't have this problem
when I kept sheep in London – three Mule lambs
in the outside loo (not in use, since you ask) – in
about 1969. Oh, and they were called Hop, Skip
and Jump.*

LAMBING GETS STARTED on a beautiful, spring-wanting-to-be-summer day. High billowing clouds, bright, warm sunshine and a light breeze just to prevent us getting overheated and overexcited. A fine single appears, just below the brow of the hill. No worries. It – it is a he by the way, I can see his little balls dangling between his slightly splayed hind legs – stands up, totters about a bit and finds the milk bar with the kind of unerring initiative that spotty teenagers demonstrate when in search of Ronald McDonald's wares.

That's all right then.

Have you ever tried moving six ewes, one hogg, one tup, several lambs *in utero*, three swallows and a rabbit with two collie dogs of pensionable age, one of whom is a bit deaf while the other would rather be watching reruns of *Animal Hospital* on telly? No? I thought as much. I wish I hadn't either.

The sheep were bored, and once bored they get wanderlust. I am only surprised that they have not taken to surfing the internet for last minute deals, so great is their desire to be somewhere other than where they ought to be. On the morning shepherding round, I found them all grazing in the wood. They were meant to be not in the wood but in the field beyond. The trouble is that once they start mooching with intent, there are only two options. You can either tie their legs together and rub your hands with victorious, if vindictive, glee or go with the flow and let them move on – though not, obviously, to pasture of their own choosing. That would be too much like allowing them the upper hand and most unwise; it is vital to stay one step ahead in the tense psychological battle that is modern shepherding.

That's why I was up there with the dogs, moving sheep and lambs – especially lambs – this morning.

The next lamb was born to Curly Horns. She had no idea what it was, what caused it or, more important, what she should do about it, so, like the good primitive sheep she is, she gave it a jolly good hammering. So violent was the assault that the lamb would have been considerably safer in downtown Baghdad in March than sitting in a field in Eden with its parent. Overnight, though, aggression turned to tedium and finally Curly Horns had a change of heart; now she protects the lamb as fiercely as she attacked it at the start.

Spikes (if you think Spikes is a silly name for a sheep, you should see her horns) was next, giving birth to a live lamb which was as dead as a stone by next morning. The next ewe had twins but announced that she was very sorry and all that but she was genetically programmed for singles, could not possibly cope with twins, and dumped the female of the pair on me. This little creature was very black indeed save for a patch of light fur over her left eyebrow. In complete accord with the Manx breed police who frown on little light patches, my husband intervened: 'Don't give her a name, whatever you do, as we shall, in the fullness of time, need to eat her.'

So, allow me to introduce Tinkerbelle, Tinker for short.

Tinker learnt to walk on a lead a great deal more swiftly than most dogs and enjoyed stretching her legs with Gyp and Tess. I probably ought to have had some promissory note, licence or contract from DEFRA for these little perambulations but I didn't. I suppose I could have put her on the Lamb Bank but black ones are never popular as it is reputedly harder to con ewes who have probably had snowy white lambkins into taking them. Discriminatory? Yes, probably, but try explaining the nuances of that to sheep.

Taking a lamb for a walk on a lead is not as daft as it sounds. At the risk of sounding like my old hockey mistress, I reckon that fresh air and exercise are as vital for sheep as for humans, and she wasn't going to get either in the solitary, confined and twilight world of a straw bale pen. And as the newborn Tinker was so tiny she could have slid down a crack in the pavement, never mind squeezed through the holes in a pig netting fence, putting her in the sheep pens was as good as giving her a gilt-edged invitation to disappear into the wide blue yonder – or even Croglin Fell – so that wasn't on either. Whilst on the subject, people often ask if I like Croglin. I tell them I don't know; I've never Croggled.

Even when Tinker was strong enough to live outside, elaborate precautions had to be taken to secure her boundaries of pen fence and gate. A brief inventory of the essential equipment runs thus: several lengths of baler twine (pink), a pair of secateurs and some green plastic chainlink stuff (that supplied in 1975 by Marley's of Walton-on-Thames is best). Snip to fit between the bottom two bars of the hurdle, tie tightly in place with the twine and *voilà*: a lamb-proof gate. Now turn the sheep troughs upside down to prevent the little devil from limbo-dancing under the pig netting, throw the lamb into the pen and wait. Does she escape?

No, of course she doesn't. Not while I'm looking anyway.

Proud of a job well done (and, more important, for

free) I stood back to admire the eclectic masterpiece of plastic horror and cat's cradle knotwork.

It was then, and only then, that I remembered the piece of retired greenhouse staging that would have been quicker, tidier, and a good deal more effective.

Tinker is now too big to squeeze through gaps but upwardly mobile enough to leap over the gate, so my problems are by no means over, just different. She has stumpy, cylindrical horns like a pair of unexploded fireworks strapped to her head and she is well on to solids, savaging Bob's Mix with all the lip-smacking enthusiasm of a Rottweiler tucking into a burglar's leg.

Of course, with such a small flock, I feel a bit of a fraud when I discuss lambing progress locally. A typical exchange goes like this:

'Has Jim/Tim/Tarquin (there are precious few shepherds called Tarquin prowling the Cumbrian up-lands but that is to cavil) finished lambing?'

'Almost. He's just got a hundred and eighty to go. How about you?'

'I've still got two.'

There's then a very long and very pregnant pause to see if I add 'hundred' after the 'two' but of course I don't. Instead, I make a fine and detailed study of my toecaps, the way I used to when I was six and had just been told off for getting ink in my hair.

The *Daily Mail* has me as centrefold today. Alan has produced the most incredible photograph of me with Tess looking pretty sophisticated in a rural, casual kind of way. I do not think I can possibly be this good-looking. Has he Photoshopped me? If he has, I don't care because I am so pleased with the results.

Susan rings in the evening. 'Dahling,' she says in her breathy, forty-fags-a-day voice, 'you're the only friend I've got who's done a centrefold without getting her tits out.' You can always rely on Susan to lower the tone of anything that remotely threatens to be aesthetically pleasing.

Lyn, who used to sit next to me in Mixed Infants, emails and says I haven't changed a bit. Oh I have, Lyn, I have. The pigtails have gone, and so have the freckles. And I'm a little bit taller too. It's a lovely thought, though.

Alan has sent me some of the other shots he took during the day and I am equally happy with these. One looks as if it should be in Alan's more usual milieu of *Hello!* and *OK* and be entitled 'The duchess at home with her dogs'. Another one of Blossom and Bob is just as stunning. Blossom looks the success story she is: rescue wreck to national newspaper pin-up in the space of a couple of years. And then some. A contented, happy, well-fed, much-loved pony. Daft, I know, but this makes me a bit misty-eyed.

What do peasants do on a day off? More peasanting, of course. The finest way of peasanting is to horse around and this weekend provided two fine opportunities.

The time has come for Mickey to have his first cross-country practice, so on Saturday Sarah hitches a ride with a pal of hers and his horse and they all set off for

Cargo. 'Lots of ditches at Cargo,' she says with mildly unnerving relish.

They arrive late as Mickey has been a little devil to box. Most horses will follow another into a dubious space, gaining confidence from the trailblazer, but Mickey, convinced that if Jasper was already in the lorry there could not possibly be room for him too, only consents to load when he is allowed to go in first. That's my boy: quirky to the point of oddness. Perhaps there is something in the Rowfoot water.

Once amongst the cross-country obstacles, though, he neither freaks nor fools, just goes and warms up quietly, taking the undulations in his stride and popping a few solid practice fences without touching a twig. How do horses know they are solid? Then on to the ditches and he's a bit of a thinker, is Mickey, so he stands and thinks a bit before realising that he is meant to do the same about a hole in the ground as an obstacle rising from it – jump to the other side of it. Sarah is worried about the green slime lurking in the depths of the ditch on the far side – well, it would make a mess of her breeches – but Mickey gains confidence as time progresses. He bounds over the solid obstacles, leaps the holes and loads up with only token resistance, farting about on the ramp once or twice but deciding that the game is up as soon as he spies the lunge whip. Then he goes back to Blackdyke to dream about Badminton.

If it's the first Sunday in April, it's Brougham Horse Trials. That's Brougham pronounced 'broom'. We have a tradition for odd pronunciation in Cumbria; it's why Torpenhow is pronounced not tor-pen-how but Trappenna. I know. I can't work that one out either. But irrespective of whether it's Brow-am or Broom we are there fence judging. This does not mean that we say, 'It's a very nice fence, can I go home now please?' It means that we sit and watch as seemingly hundreds of competitors flow over jump number 18. We have to record their passage, noting whether they run out, refuse or reverse, whether the rider physically abuses the horse or, since we live in sensitive times, swears at it or us, whether either component of the combination falls. If the rider falls off, we scrape up the bits, and if they swear, we are instructed to note down what they say. To join the brotherhood of fence judges, it is vital to possess a comprehensive knowledge of swearing in order to be competent to assess the gravity of such transgressions.

There we sit, fuelling ourselves with coffee and, later on, spicy lentil soup to stave off hypothermia as competitors, anxious parents, trainers, minders, grooms and dogs walk the course before tackling it. The horses, of course, enjoy no such privilege. That would be altogether too risky: some might decide to pack up and go home before they went down to the start gate.

Gyp and Tess snacked on a Jumbone for breakfast at nine thirty as they had forgone the normal rations at home in favour of Stugeron to get them here. They were not given an option, obviously. Now, lashed to the back bumper of the car, they settle down for a deeply social day, paying scant attention to the course walkers and far more to their accompanying dogs. Jack Russells on

Queen Anne legs, loping lurchers and sleek black Labs predominate.

G&T love horse trials; one meets such a nice class of human. Tess did not wash for weeks after Armathwaite Hall trials some years back when Mark Phillips patted her and since then, though the friends they have made have not been as luminous as the Captain, they have always been none the less delighted to go out and mingle with their public.

A headscarfed, Chanel-lipsticked, Hunter-wellied aristo throws Tess a glance. 'Looks just like that dog that was in the *Mail* yesterday, Henry,' she says. Henry's not listening. He's inspecting the rosemary bushes in the tubs on the groundline of Fence 18. Hands off, Henry, those are mine at the end of the day. The chief course builder has said we can take them home: the wages of fence judging are botanical, you see.

It is a long day from the briefing at eight thirty in the morning through to nearly seven in the evening, with a snatched quarter of an hour lapse in the proceedings for a pee-break and lunch. But hey, someone used to do it so that I could go out to play on Sundays, so it seems only fair that I do my share now that I'm too old, too sensible and too broke to participate any more.

Back at Rowfoot, we had missed the window cleaner and the Jehovah's Witnesses, though the latter had most thoughtfully pushed a leaflet through the door so that we can make contact with them if we wish. Next year we could well have a runner at Brougham, and if that's the case I shall need all the prayers I can get, so I put their leaflet carefully to one side.

*   *   *

No lambs next day, nor the next. Mind you, with only six in lamb this time, gaps 'twixt births are to be expected.

Wednesday and it's freezing. Where did that promise of summer evaporate to? If Mother Nature can't make up her mind what season it is, she should clear off to Mars – everyone seems keen to find life up there at the moment but there won't be much surviving down here if this horizontal rain carries on.

Now it's turned to horizontal hail.

And out in the field, wouldn't you just know it, there is a pair of rather cold twins and a potentially distressed parent. The twins are too far apart for my liking. She could lose concentration and forget to bond with both of them if the weather blunts her instincts, so I decide to intervene. Pick up cold infant one, then cold infant two, and set off across the field without – take note, because this bit is important – making eye contact with mother. It's then that she decides that nothing on earth – even horizontal hail – will blunt her maternal instincts, that I am the childsnatcher from Hell and that she will do her own bit of intervening.

She creeps up behind me, Grandma's footsteps style, and clouts me in the back of the knees. I sprawl in the mud, dropping, to her immense satisfaction, both of the lambs. Being an opportunistic sort of sheep, she takes full advantage of my indisposition and bashes me on the right shoulder with her horns. Always an optimist, I decide this could have been worse: it might have been my head. It could be wise to wear my jockey skullcap if I get involved with moving sheep in mud in future, I think, crawling to my feet, wiping the mud from my hands down my sodden jeans and picking up the lambs again. I glare at the woolly old bag: 'I don't give up that easily.' Why am I arguing with a sheep? Is this the first sign of some obscure dementia? Is there a name for it, or a cure?

We make slow progress, as now I glance down to check she's still with me. After all, the lambs now smell of me. And she won't like that one bit.

We reach the gate. I tuck one lamb under my left arm, open the gate and shuffle through. Thankfully, all the other ewes just stand and gawp but make no attempt to follow. I shut the gate, turn round and edge forward. She gets me again. Down I go, lambs lobbed right and left. This time, though, I know that she knows my intentions are good. How do I work this out? She doesn't attack me this time, that's why. For a Manx ewe, that's remarkable gratitude.

In spite of my intervention, she has remained close to the lambs all the way down the field, her nose rarely leaving one or other swinging body. Occasionally, she shoots out a pink tongue to lick briefly, and constantly emits low bleating, reassuring them and terrifying me.

Good mothering is what you need in breeding ewes but not always what you get. With Manxes, the story is just a little more convoluted. Usually brilliant mothers in their natural homeland on the hills and rocky out-crops of the Isle of Man, they are unlikely to 'do' so well that they carry and produce twins. Consequently, some ewes tend to assume that they are genetically programmed to have singles and when another lamb appears their reaction is occasionally one of confusion. Sometimes confusion turns to antipathy and then to rejection. I have seen a ewe tend and lick one lamb and unashamedly kill the other if she decides she is that way inclined. So, you see, I am very grateful if she loves both of them.

Two days in the warmth of a dry stable and the ewe and her twins are ready for the great outdoors again.

Off they go, trotting merrily up the field, a happy single-parent family.

Sunday and one of the single lambs is missing. The twins and their mother are there, set slightly apart – as normal – from the others, but there is no sign whatsoever of the single. Worryingly, its mother comes to the trough along with the rest of the flock betraying no suggestion of having lost, or even temporarily mislaid, her lamb. Very odd indeed. I fall to pessimism now: perhaps she knows something I don't. If, say, she had seen it borne away by a fox, she would not bother looking for it quite simply because she would know exactly where it went: between the molars and into the small intestine of Reynard.

I prowl around a bit, scanning the grass for a small, inert black heap. Nothing. I walk the boundary of the lambing field, checking dykes, rabbit holes (that's happened before – a lamb sneaks down a rabbit hole to get out of the draught and, once there, can neither turn

Background: The view from the farm gate on a frosty morning.

Right: Home.

Bottom: The Eden Valley, an easy place to love and a hard one to leave.

THE MANY FACES OF KATIE MORAG.

This is Katie's 'it wasn't me' look. She wears it quite a lot. And if you can get cross with a dog with a face like that, you have a heart of pure titanium.

Right: Catwalk Katie. Elegant, graceful, and in some places completely transparent, so thin is her skin.

Below left: She corners like a Ferrari. It's probably a good thing that the 30mph speed limit through the village doesn't apply in the fields.

Above right: Artist's model. Immortalized in pastels by my talented friend Jane Fay, whose uncanny ability to catch the very essence of her subject is somewhere between laudable and downright spooky.

Right: Anyone seen my hair gel?

Background: Rajah, twenty-eight years young. Retirement put quite a spring in the old boy's step and a sense of mischief in his heart.

Right: Having been on the Hay Diet all winter, Blossom is glad when summer comes around and she can try the Grass Diet instead. Neither makes her any thinner, incidentally.

Right: Three's a crowd. Bob cannot bear to look as the love of his life cosies up to Arthur; theirs is a very peculiar relationship indeed.

Left: Blossom – another of Jane's lovely pastels.

Top left: I often think that John ought to get danger money, and probably protective goggles, when he shears Manxes.

Top right: Arthur, Rowfoot's very own Monarch of the Glen.

Left: G&T yoked together and deep in sisterly conversation, when really they should be fence-judging.

Background: Flying pigs make happy pork.

Top left: The flock. The breeding ewes, present and future, along with Chops – the Next Generation.

Top right: Tinkerbelle hasn't yet grasped that it is bad form to dribble in public. Her collar and lead, ready for walkies, are just visible.

Below: My namesake, the Highland calf Jackie. I have serious doubts that she can see where she's going.

Are we by any chance related? My best boy, Mickey.

round nor find reverse gear), behind tree trunks, in corners. Nothing. I walk the perimeter of the wood, then round the long meadow and round the lambing field once again. Still nothing. And mother is still guzzling her way through Bob's Mix with the others, oblivious.

It seems a good idea to check the road side of the hedge; a lamb might have somehow squeezed through the wire and then got snarled up in the bushes. Nothing there either.

Baffled, I go back indoors and explain to Malcolm who quite clearly doesn't think I have looked nearly carefully enough. We set out together and do it all over again. Now he believes me. The only possible explanation is the fox theory. Or perhaps it's been rustled: yeah, right. Rustlers are always on the lookout for a spindly black lamb of indeterminate origin. Worth a fortune, them.

The words 'Ah, booger' spring to mind and, deciding that there really is nothing more to be gained by loitering in hedges and patrolling boundaries, we go to get the Sunday papers. But we drive e-x-t-r-e-m-e-l-y slowly . . . and peer in the hedge again. We are, you see, in denial about the disappearance. Lambs, I suggest, do not dematerialise like something off *Doctor Who*. They have to go somewhere.

We come home, clutching copies of sufficient Sunday reading matter to last until next Friday week, and creep along the hedgerow again. Just in case.

Monday is Bank Holiday and that is a good day to stop at home and, if you get truly bored, look for lost sheep. I get bored. I look some more.

*　　　*　　　*

On Tuesday we fly to London from Newcastle after a quite unnerving experience with the airline tickets. We don't have any. Well, you don't these days. It's extremely disconcerting, turning up at an airport in the confident expectation that a machine or an official will fail to recognise you and it doesn't happen. Instead, a very nice girl in a BA uniform persuades me – against my better judgement – to try the automatic check-in. Now, I'm no Luddite and I'm all for modern gizmo-wizardry, but how does the little man in the guts of the machine know that it's me with this credit card? My name comes up in lights, our boarding passes spill out into the document tray, we are told to lob our cases in at Desk 18 and be on our way. This is all fine, except for the fact that the people in the long queue stretching from the entrance to Desks 1–17 inclusive all think we have jumped the queue – their queue – and scowl murderously at us.

I don't care. I've lost a lamb and in consideration of that alone I deserve compassion, understanding and possibly compensation. Squillions of the stuff. And an unencumbered queue.

London passes in a haze of BBC radio interviews and a champagne lunch; our return flight is similarly queue and hassle free and the hills of home beckon as enticingly as ever. Going away is good for you: coming back is better. It reminds me that I live in the most beautiful, favoured corner of England.

*　　　*　　　*

I ring Joan to say that we are home but she's out. As my mother used to say, the cheapest time to phone your friends is when they are out.

Joan rings back within minutes. 'I was outside, busy with a lamb.'

'I thought . . .' What I really thought was that this time last year, Joan explained to Freddie in words of one syllable that she was not, ever, going to mess on with pet lambs again for him, and if he wanted to rear pet lambs he would have to do it himself without her help. So I thought but I didn't say.

'Your lamb. It turned up, see.'

Freddie arrived in the morning to explain all.

'Where is it, then?' I asked. 'Is it dead yet?'

'No it bloody isn't deeed. It's in the boot of the car, in a bag. Though it's probably not in its bag now, but peeing everywhere. Still, he can't do much damage in that old thing,' said Freddie philosophically.

And the rest is Freddie's story, not mine at all. After all, I wasn't here. I was in London, marvelling at what Ken has achieved with his Congestion Charge and wittering on in the bowels of BBC Broadcasting House.

'Someone found it wandering on the road and took it down to Ainstable Hall. They took one look and said, "Snot ours"' – I can't help it, that *is* what he said – ' "'tis Moffat's."' No one seemed terribly clear on how the next bit came about but Freddie was summoned to collect it. 'Its eyes was all stuck together. Matted. And it couldn't stand, it was that weak.' I think Freddie had a few doubts about the 'wandering' bit of the description: 'staggering' might have been more appropriate.

But he took it home anyway. 'Jean gave me some cow colostrum for it and it had a good soock on it, so I thought it was worth a go.'

Freddie's younger son did not share his optimism. The lamb looked nearer dead than alive to John.

'What ya gonna do wid it, father? Clout it on the head with back of a shovel?'

Freddie, rightly, was appalled, but there are plenty of farmers who would think nothing of doing just that. But Freddie thinks that if they want to drink they deserve a chance. Amen to that, say I. And so would you, if you had seen him: a jet black lamb, hopping and bouncing with life. And piddling. And not in a bag. Freddie was understandably thrilled to have pulled the tiny Manx back from the brink, though where he spent the days between Sunday when he disappeared and Wednesday when he turned up, dehydrated, cold and starving on the road, remains a mystery.

'Only he knows and he's not tellin'. And his mother didn't want him when I took him back up the field – denied all knowledge, she did.'

That's the Manxes we know and love. Once the bond is severed it cannot be repaired with Manx or probably any other primitive sheep. You might just be able to persuade a mule that her long-lost offspring has, like the prodigal, returned but Manxes are far too intelligent to be conned like that.

Oh, and the lamb's name? I've called him Lazarus, since you ask. It seemed appropriate.

*　　　*　　　*

Early morning and the shepherding round. I'm not a morning person, but even I can manage to count to seven just after the hour has struck. Except that I can't seem to get to seven. Five, yes, but not seven, the last ewe having lambed twins with so little drama that they didn't even qualify for a diary mention.

One, the single; three when you add the big twins, who are now old enough to be chasing each other up and down the dykes, even in pouring rain. I can get to five when I add the second set of twins, and that is it. Two are missing. But hang on. That ewe at the top of the field under the oak tree is not coming to the trough. Not much puts Manxes off their grub, so I inch closer for a look. She's sitting regally, staring at me but still not shifting. Just downwind of her is a vast tree root with a bit of a hole behind it: ha! Two beady little eyes stare out from the woody cave – a lamb. I kneel down and stick my fingers in the hole, feel around for its front legs and haul gently. It wriggles towards me, snagging a bit on a twiggy strand of rootlet, but out, anyway. Another pair of beady little eyes peer at me quizzically; but the owner of this pair won't be dislodged. Obviously, these tiny lambs found the nocturnal hailstorm scary, cold and very, very wet so they sensibly sought refuge amongst the rooty fingers of an ancient oak. They should have been more careful with their feet, though, because it appears that twigs and leg have become entangled. As I've got leggings on I kneel and scrabble away some soil, more soil, more soil still, until I can unfold the spindly limb and disengage it from the root system. A slightly dusty lamb totters out and makes off to mum for a drink.

Next morning, I can see all seven clearly enough. The trouble is that two of the little blighters are pulling rude faces at me from the wrong side of the fence. This is typical of sheep: give them a nice field full of luscious grass and what do they do? Tunnel out into a bare expanse of plough, upsetting rabbits as they go. I'm with them on that, though, as there are far too many rabbits about: it is time to ring Thomas and his Murder Squad friends again. These lads don't investigate murders, they commit them, and with luck you or your dog can eat the victims.

I chase the escapees down to the stile, where I am fairly sure they have got through, being just small enough to squeeze through under the bottom rail. One gets stuck in the thorn bush. I try – God knows why – to reach him and extricate him. I grab a back leg. He contorts violently and my wrist tears on thorns. What the thorns miss, the barbed wire gets. My hand is shredded and bloody. I do not like blood and curse myself for leaving the aluminium crook behind. I have a beautiful new hand-carved crook, a very precious gift from stick-maker Jim Landles of Northumberland. Jim has worked the head of a Manx tup into the horn but there is no way that one ever uses an object of such beauty to catch sheep. An aluminium one that bears the scars of fruitless attempts to nobble sheep in the past, with more dents, dings and bends in it than a banger-racer, that's what you need for chasing lambs.

Finally, the remaining lamb dances happily down the dyke and shimmies through the base of the stile to join his pals. I shout stupidly after him that one day I shall have the last laugh – unless he dies first, of course.

I must stop shouting at sheep. It is not normal behaviour for a grown-up.

The family arrive for the weekend. It occurs to me that there is merit in making goddaughters compulsory. Just when you feel jaded and cheesed off and go into full shouting-at-the-television mode, they bring undiluted sunshine and remind me how much I like children. After all, I was one myself once. If you're especially lucky, goddaughters save small morsels of special delicacies for you – cornflakes, for example – ready masticated and stored in their teeth. They are very thoughtful like that. But they are equally capable of compromising the entire family's reputation, par- ticularly during those school role-play sessions, during which teachers uncover dark family secrets. When all the other little girls went along the usual sit-down-mummy- and-I'll-get-you-a-nice-cup-of-tea route, one of my pair piped up: 'Sit down mummy and I'll get you a nice glass of red wine.' We're probably all branded alcoholics now.

Ellie walks rather than gallops on walks these days – that's what growing up does for you – but it is fruitless trying to wear Phoebe out: she is possessed of endless energy. We take the dogs out for a walk and Phoebe hitches up her leggings above her knobbly little knees to reveal white Olive Oyl legs. Atop this pair of unlikely limbs she canters coltishly along the grass verges in the village, threatening to paddle in streams, climb on walls and leap from parapets until even the dogs are exhausted. Finally, back at home we plunge her and her quieter sister into a Jacuzzi bath where they yelp with almost-exhausted glee.

I just know that I have years ahead punctuated with dissonant pop music belted out by youths with odd body piercings, unsuitable boyfriends, absurd clothes and green hair episodes. Still, who knows, with a pot of ink, a little imagination and a pliant third party, they and I could create a full set of traffic lights.

The next day Sara and Malcolm (her Malcolm, not mine) and the two girls go off to Beamish for the day. They will love it. We stay at home to watch the sheep and, as the day develops into a sunny one, we fire up the barbecue for a homecoming feast. The wanderers return towards evening.

'Did you have a lovely time?'

Phoebe, her face a complex mix of enchantment and exhaustion, replies, prefacing the word ''ot' with another, suggesting that a family with a firm grasp of Anglo-Saxon meteorological terms had been within earshot during the course of their lovely warm day at Beamish.

She's coming along nicely. Probably wiser never to let her go into the world of horse trials, though; with a vocabulary like that she could get into awful trouble with fence judges.

*   *   *

One evening, when I thought lambing was over, I watched them contentedly, thinking that if I didn't know better I'd have said Spikes looked like lambing. She couldn't, of course, as she had lambed three weeks ago. Next morning, Spikes had two lambs, the size of guinea pigs, with drinking-straw legs. They died, of course.

May

Ours is a spring garden. It is at its best now as papery pink poppies flutter in the breeze under the early summer sun. The fragile flowers are beautiful, but their origins are vague; their petals are palest pink, their centres velvety ebony. All the poppies in this and the neighbouring village appear to have descended, by devious routes, from the stock of one elderly spinster's garden but no-one apparently knows what variety they are. I suspect opium, myself.

MOST OF THE sheep are looking after themselves now. The grass in the top meadow is rich and plentiful, the field having been left unused throughout the whole winter. It is also worm-free, an added bonus with the babies. The tups roam about with the ponies, though the eldest, Arfur, has developed an extraordinary relationship with Blossom. They are close. Unnaturally close at times. Take this morning: Malcolm shrieked from upstairs, 'Quick, get the camera.' So I did, because after twenty-eight years of wedded something-or-other I have learnt to do as I'm told first and negotiate or possibly argue the toss later on.

It was worth it. Or it would be if Photoshop hadn't been invented because no-one, surely, would believe that the pictures were anything other than digital trickery. Bob standing guard as usual, Blossom lying down, sphinx-like, and the tup snuggled in to Blossom's meaty flank, his monstrous antlers curled over her back. I snapped off several, taking the risk of shooting into the sun and hoping they would come out: they did, but I still don't expect anyone to believe they are not spoofs. Arfur clearly thinks he's some sort of honorary horse; he hangs out with Blossom and Bob, trailing in their wake and occasionally quibbling with Bob over their place in Blossom's affections. Once, I saw him run at Bob, head down in a determined assault: pick on someone your own size, I thought.

The pet lamb Lazarus spends his days in a hastily constructed pen, made up of some old greenhouse staging and two hurdles, held together with Farmer's Friend: cerise baler twine. At night I return him to the safety of Mickey's old stable, in a pen of stacked straw bales. Unlike the politicians in Iraq, Lazarus has an exit strategy: he jumps out. So the number of bales increases and now he is in a straw pit with sides three bales high. The whole edifice has to be demolished morning and night, and then rebuilt; this fails to deter him and he makes valiant attempts to scale his perimeter walls. I shall be glad when he's outside and all this construction and demolition work can stop. Until then, it's all a bit of a fiddle, but at least it means that he can enjoy the sunshine on his little black back instead of sitting in gloomy semi-darkness.

Each morning, he scoots up the yard, jumping madly left and right, with the dogs in licky attendance. Gyp tends to take care of his bottom, Tess sees to his eyes and mouth, and between them they add up to one half-decent, loving mother. The loving mother herself appears hugely gruntled at having rid herself of her pesky offspring, instead of having her teats systematically bashed, her back stood on and her peace and quiet disrupted like all the other ewes in the flock. She just smirks knowingly.

I could kill her.

Lazarus drinks with fierce enthusiasm; his lamb milk substitute costs roughly double what he is actually worth, but who's counting?

As the weeks pass, he develops a ritual during feeding time. While his little black rubber lips suck furiously from the red teat screwed on to a Volvic bottle, his back legs perform a manic dance, pounding like a pair of perfectly matched miniature pneumatic drills. Every so often, and for no apparent reason, he tap-tap-taps his left front hoof too. All this frantic activity should give him chronic indigestion but it doesn't. As soon as he's finished, and his tenacious grip on the teat is released, his batteries run out abruptly; the dancing stops and he farts melodically by way of thanks.

I do like him: he's a great little chap.

The diary for May looks slightly bonkers. During the first week, I do a number of book talks and then unexpectedly I get a call asking if I can get to London on

Monday – this is two days off – for the Johnnie Walker show. Can I get there? You bet I can.

I book the last couple of seats on a Sunday flight, ring to see if our favourite hotel can accommodate us (adding that on this occasion I would happily settle for the broom cupboard), sort out the babysitting arrangements for the lamb and start looking forward to the whole thing enormously.

There are myriad rules about moving sheep these days; I do not follow any of them. The paperwork would take far too long, and since the sheep in question is only one size bigger than the average domestic cat it all seems ridiculously over the top. Lazarus has a temporary billet and no, I'm not saying where.

I stick him into a Laura Ashley carrier bag with his head poking out between the pink handles and deliver him and his rations to a secret location.

Sunday morning, and before we go jetting off to London it's time to go pony trekking. This is Poppy's birthday treat; Poppy cycles up to Rowfoot every Saturday and helps me groom the ponies. A trekking treat seemed a better idea than a wrapped-up sort of present so off we go.

The ponies are sturdy, sensible types. There are a number of solid, softly coloured Highlands, a smattering of hefty black and whites, several smaller Welsh type ponies and then there's Sugar. Sugar is twenty-three and has been there, done it and had T-shirts been invented when she started out she'd have a whole wardrobe full. Poppy rides Sugar and I am allotted Monty, one of the Highlands. We set out through the back gate – oh, the joy of no road work! – and up on to the fell above Ullswater. Before we set out, Poppy had offered to groom her pony, but the owner tells us, 'I'm afraid I can't let her. Sadly, children under fourteen aren't covered by my insurance . . .' And the whole insurance madness gets even dafter when I ask our escort what she has in her saddle bag. 'Well,' she replies, 'it's a first aid kit. The trouble is that none of it is much use as we aren't allowed by Health and Safety to put plasters on kids because of the possibility of abuse or on adults because they might suffer an allergic reaction and then they could sue . . . Mind you, I suppose the water bottle might come in handy.' She ponders a bit and adds brightly, 'If someone fell off and broke a leg or two, it was scorching hot and the air ambulance was delayed by four hours' – I am struggling with this implausible scenario – 'at least they wouldn't dehydrate.'

What a barmy world we have created. And how sad that no suburban ten-year-olds can enjoy that special freedom of pedalling furiously through Richmond Park at dawn, spending a long day mucking out and grooming holy terror ponies, and then cycling back alone in the fading light without fear of muggers, paedophiles, insurance claims or allergic reactions.

The sun shimmers on the lake below, reflections on the ripples in the wake of a leisurely steamer glimmering like so many silvery petals scattered on the surface. There's a haze on the hills beyond; an occasional mountain biker or dog walker calls a morning greeting and above us buzzards, peewits and clouds scud across a clear blue sky.

It is very difficult to believe that in a few hours I shall be in London.

Poppy's continuous smile indicates that she is having a wonderful time; she sits easily, naturally, on Sugar, coping commendably with the changes in gradient and pace. Sugar, meanwhile, exhibits quite clear ideas about her personal space, every so often glowering at Monty and warning him to keep his distance. He does: wise boy.

Poppy's mum Wendy drives us back to Ainstable and we muse on the daftness of insurance. 'We had trouble getting insurance on our house because it is so near the beck,' she tells me. 'One bit of the application form asked whether the house had been flooded in the past five years and of course I had to say yes, because it has. Just the once. It made the insurance company very twitchy indeed. They phoned up to ask whether any effort had been made to remedy the situation since the flood and I said yes, of course it had. Then they wanted to know exactly what had been done and I told them that someone had pulled the dead sheep out of the beck . . .'

The dead sheep, you understand, had effectively dammed the flow and the water had backed up with catastrophic results. No doubt the urban insurance company was impressed by this rural problem-solving initiative.

Ullswater at 10 a.m.; London by 5 p.m. Quite a contrast.

Monday – and a day to kill. We're back at the

Howard, our chic little billet on the Embankment. It is good to stay somewhere where the plumbing is not an initiative test and the staff greet you like old friends: odd, really, that such a swish establishment manages to strike such a home-from-home note. At home in Cumbria we don't get too many Taiwanese businessmen barking into mobile phones over breakfast and the night sky is filled not with bright lights but with stars. We take full advantage of our free day, crossing the wobbly bridge to visit Shakespeare's Globe where rapt seven-year-old faces are entranced by tales of plays and groundlings, and doing the brace of Tates, Modern and Britain. I haven't had so much culture since I last ate bio-yogurt.

Eventually we come to the whole point of the trip: the Johnnie Walker show.

It is a deeply weird experience meeting someone you are convinced is a close personal friend but who, in reality, you don't know from Adam. I've listened to Johnnie ever since I was old enough to own a radio and he is easy company. 'What do you think of this shirt?' he asks. 'Sally's taking the piss . . .' We are not on air yet, you see. 'Oh, Johnnie, it's a bit Radio Caroline, isn't it?' Well, it's got little pink flowers on it and it makes me think of Scott Mackenzie tripping off to San Francisco . . .

Johnnie puts me instantly at ease: this is great fun. The team who produce and twiddle buttons are fun. So is Janine, Johnnie's researcher. She points out where Steve Wright's studio is – just over *there* – and Sally Traffic trots in and out with motoring information. I would love to be able to remember every second in fine detail for ever. I won't, of course, but that won't stop me trying.

A bloke called Steve from Portsmouth texts in while Johnnie's playing a record and says that *The Funny Farm* is the best book he has ever read. My cup runneth over and I am mightily impressed, partly because I don't even know anyone called Steve who lives in Portsmouth but mostly because texting baffles me comprehensively. Plus, if I wanted to contact the BBC from home, I'd have to ring Directory Enquiries first.

If it's Thursday it must be Birmingham. And I had been dreading this: speaking at a Saga Literary Lunch, sharing a platform with Joan Bakewell and Robert Lacey. It is difficult to work out which one of them I am more terrified of, sophisticated Joan or urbane Robert. I need not have worried. Both are friendly, funny and charming; neither makes me feel like an impostor and both laugh at my jokes. Saga make us all hugely welcome and the event seems to have been a terrific success.

Isn't life strange?

Saturday, Blossom's photographs, the 'before' ones, arrive. I would like to be left alone in a darkened room in possession of a blunt knife with the person who allowed her to get into that state. They tell me there are some fine recipes for lambs' testicles; I am sure that with just a little judicious adaptation from someone like Hugh Fearnley-Whittingstall we could come up with an appropriate dish featuring the human version.

*   *   *

Sunday and it's time to castrate the lambs. Now, no matter what the books tell you, it is just not possible to castrate a lamb until he drops his spherical appendages. It doesn't matter how many directives come out of Brussels or what DEFRA says about The Deed's having to be done within so many days of birth: if those balls aren't down, you cannot deal with them. I speak as one who knows . . .

I have just three boys this year and I have – how can I put this delicately – been checking their progress regularly. The two in the field have just let their undercarriages down but Lazarus is hanging on to his. He has maybe noticed the unpleasant glint in my eye and has worked things out for himself.

So, I move them all in readiness. Never lose your temper moving sheep, never shout, never wave your arms, threaten or holler, and they will do just as you wish. Honestly.

I open the top gate and they trot through. They stay there overnight: they are prone to overexcitement when they are on fresh pasture and they've not been on here for a fortnight so there is sufficient fresh growth to inspire naughtiness. Best left, then, until morning. Patience is a virtue you need shedloads of, if you have sheep. Next morning, call the tups, cage the horses, open the middle gate and lo, they trot through there too.

I've tied a couple of hurdles across the bottom half of the pens to make the space enticing: when I have oodles of lolly, I shall get proper hanging gates everywhere, but until then, baler twine does the trick. In they go. Gate shut: sheep captured.

Freddie arrives looking none the worse for wear and that's a surprise because it was his retirement jolly last night. 'God,' he says, 'some o' them fellas can drink. Mind, they've got the storage for it.' There's a sense of childlike wonder in this statement. We rugby tackle the little tups with a nifty bit of footwork (that's Freddie) and a hookie from me and the crook.

And now, here is a masterclass in How to Castrate a Tup. Well, you never know when this knowledge might come in handy.

One: sit the tup on his bottom. Gravity will provide a modicum of assistance, that way. Grope about a bit. Swear at him when he withdraws his balls into the region normally occupied by his neck. He does not like what you are doing to him. Tell him that twenty – oh, all right, thirty – years ago, grown men would have been very grateful for similar attention.

Two: be patient.

Three: be patient a bit longer.

Four: take the nutcrackers (well, that's what I call them – it seems appropriate), place the little rubber ring on the four spikes – think French knitting here, girls – and open the implement to its full stretch.

Five: watch the little tup's eyes cross. When they've uncrossed, proceed.

Six: take his important little furry bits in your fingers, make sure you've got both testicles and pop through the fully extended rubber ring.

Seven: check again that you still have a matching

pair, pendulous and prepared, and close the nutcrackers round them.

Eight: release the ring and express the opinion that this procedure could be usefully applied to several named individuals. When you've finished listing them – it could take a while – catch the next tup and repeat the whole ritual.

Fun, isn't it?

They may bleat with a slightly higher tone for a day or so, attempt to walk with their hind legs crossed and spend more time sitting down with their buttocks clamped together, but they will get over it.

Some days do have a subtle theme, and today it was undeniably balls. The next spheroid to demand my attention was one connected to the water supply: a ballcock. I had found it irreparably shattered and tied the arm up with Farmer's Friend – the pink sort – the day before we had gone to London. The one thing you cannot afford, if you are on a water meter as we are, is for the old H_2O to leak away, because you are paying for every drop whether you use it or lose it. It drives me to distraction and sometimes drink – though not one with water, obviously – when I see visitors running the cold tap so that their glassful will be a chilled one. Fill the kettle with the stuff, for goodness' sake. Or run it into a jug and I'll water the houseplants with it later. Or fill the dog bowl. Do anything but waste it.

So we headed off to the trough above the Blue Corner, armed with a bucket containing a wrench, a penknife, an old chunk of sponge from a dead sofabed and a shiny new orange ballcock. There was some murky silt in the base of the trough and since neither sheep nor cattle much like murky silt, it seemed a good time to clean the trough out. Now, I'm sure you'll have worked out what the wrench, the penknife and the shiny new sphere of orange plastic were for but perhaps you were wondering about the sponge from the old sofabed. There comes a point when the quantity of silt you can scoop up in the bucket is insufficient to justify the effort and this is when you use your sponge; if you are as hamfisted as I am, you too will end up with an impromptu selection of little fake beauty spots decorating your face and most of your upper body. But stick with it. Eventually, the trough appears to be clean. Or at least cleaner than it was when you started; the silt is mostly gone and what is not deposited on the grass nearby you are wearing.

It is now time to screw on the new ballcock, release the arm from its baler twine prison and let that water flow. Now you can go and collect the papers and check the trough on your way home: it will be full and in spite of your efforts with the sponge it will look filthy but fear not – it should settle. Just be patient, the way you were with the sheep's dangly bits.

Job done. Now get yourself a monstrous G&T and sit on your swing bench in the sunshine because, after all, it is Sunday.

Monday and I have an appointment at our doctor's surgery. I am not ill but I can't shake off this cold. It has hung around a long time and been harder to shift than an unwelcome relative at Christmas, so I have been summoned by the practice nurse who had heard me sniffling unhealthily while collecting a repeat prescription. 'We'll do a spirometry test,' she trills happily. Spirometry? That sounds suspiciously like something mathematical that I couldn't master at school. She is concerned that my inner tubes have been damaged beyond the attentions of Kwik Fit by the pyres that raged during that hideous summer of 2001; it can't be anything else because I have never smoked (though there was that bit of wicker chair when I was six; I had thought it must be like the weed my parents smoked but fortunately it served as powerful aversion therapy rather than a catalyst for lifelong nicotine addiction) and I have never shoved anything more invasive than an investigatory finger up my nose.

Spirometry over, I'm packed off for a chest X-ray. Bloody pyres, still wreaking havoc three years on.

Diversification is a many-splendoured thing. You acquire new skills, you make a shilling or two and, best of all, you make new friends. And if you are very, very lucky your new friends might even start growing pigs and driving up the motorway on a Saturday night to share their sausage with you. Don't look so disbelieving: that's exactly what Justine and Paul did.

Justine and Paul had a weekend in the Old Dairy Cottage last year for Paul's birthday and we have kept in touch via email since, recognising in each other a kindred Mad As A Box Of Frogs sort of spirit. Talking of email, Justine's address starts with her alias 'dizzymare', which tells you much that you need to know about her.

Justine likes to take Paul somewhere in beautiful countryside for his birthday (we were an obvious choice, in Eden) where they can pick people's brains (bit of a struggle with that component of the Master Plan) because they are that rare thing: Wannabe Farmers. Furthermore, they want neither handouts nor subsidies; in return for meagre financial rewards they simply want a sustainable country life, alongside animals they care for a great deal more assiduously than many people do their own children.

As others leave the land and sell up and out to foreign exchange dealers and IT consultants, Jus and Paul are still quarrying out their farming dreams from the unpromising rocks of putative profitability on their rented smallholding, Tawd Moss.

They are no strangers to stockmanship; Paul managed a successful commercial thoroughbred stud before turning his attentions to pigs, and they have reared turkeys and ducks for Christmas alongside their day jobs. ('Never again, ducks,' says Justine – I think it has something to do with escaped poultry, a wet night and a ditch. Best not go there, really.)

From the quality of the meat he and Jus are pro-ducing at Tawd Moss, it is evident that Paul is now translating the principles of husbandry that resulted in such healthy equines to pig management. There's a kind of sixth sense that a good stockman possesses, that ability to walk past an animal and know, just know instinctively, whether all is well or not: something in the coat, or the brightness of the eye, or the attitude, or something so small and so subtle that it is invisible to the naked eye. That sixth sense is the mark of a real stockman and you've either got it or you haven't: as a combo, Paul and Justine undeniably have. In spades.

Of course, any fool can keep livestock. And some do. Even if you don't know what you are doing when you start, you owe it to your animals to learn Pretty Damned Quick; to be alive to the gaps in your knowledge and, most important, to recognise those occasions when you are totally out of your depth. As that noble Latin proverb says, 'Semper in excreta sumus, solum profundum variat.' (We are always in the do-dos, it's only the depth that varies.) When things get that bad and you really don't know what you are doing, you need to find a man who does. Fast. It might not be a bad idea if some certificate of competence were required before anyone could go out and buy animals, to protect unwitting livestock from dimwitted owners. The system could never be extended to humans, though, without first-time parenting's becoming a job no-one would get: no qualifications and no previous experience either.

I ring Jus to see how the day is panning out, and whether I can still fit in a hack on Blossom – it's a fine, sunny morning with high clouds and a fresh breeze.

'Yes, we'll leave about one thirty and be with you later in the afternoon.'

'I thought we were meeting at Junction 37 and you were going off to look at Gloucester Old Spots in Sedbergh?'

'No. I told Paul there was no point because we

haven't got anywhere to put more pigs at the moment. So he's just come in and said, "You know, Jus, I don't think there's any point in looking at those Gloucester Old Spots because we haven't anywhere to put more pigs at the moment," and now he thinks it's his decision and we're definitely not going.'

Women everywhere will identify with this.

They arrive at half three and that's exactly the right time for an adjournment for tea and an update on developments at Tawd Moss.

Their first foray into pig farming has been with Oxford Sandy and Blacks, reared outdoors on the principle that happy pigs make the best pork. Itchy, Scratchy, Dotty and Big Frank have ingested neither a cocktail of sinister drugs nor any of their distant relatives, just an additive-free mix supplemented by the occasional Scotch pancake and a milky drink at teatime. (I once heard of pigs raised on a diet of jelly babies and bourbon biscuits because both commodities were a) stale and therefore b) cheap, but milk is neither and definitely best.)

Although Europe banned iniquitous growth pro-moters during the late nineties, the use of prophylactic antibiotics remains lamentably widespread. In Denmark and Sweden, antibiotic usage has actually increased since growth promoters were banned and since most of the pigs in those countries are kept indoors (been to Denmark? Ever seen a pig? Thought not) this suggests that outdoor-reared animals are healthier.

Even here in Britain, there is widespread concern within environmental groups about the escalating use of drugs that make pigs grow faster. Part of the problem here is the resourcefulness of bugs in staying ahead of the game: it's a case of more drugs, more bugs and, what's more, bigger, stronger, nastier and increasingly resistant ones. So if you don't want to eat factory-reared pigs, start asking your butcher's shop, restaurant, or hotel exactly where they buy their pigs and how they are reared. You never know; you might be avoiding serious damage to your health. Someone bothers to wag a finger at you from the back of a fag packet about the deleterious effects of all that smoke and gunk on your lungs, but you have to use your own initiative where rashers of bacon and your innards are concerned. So be sure to ask where meat has come from and how it has been reared, for we are, as you know, what we eat. And if you would prefer not to be a walking chemistry set, then contact someone like Justine and Paul, who do bother about these things on your behalf.

The pig that Jus and Paul brought to Rowfoot was a bit of Big Frank, named in honour of their pigging mentor, er, Big Frank.

The deed had been done the previous week at a small local family-run abattoir. Paul says he has gone right off the word 'slaughterhouse' since they've kept the pigs and I am with him on this. The pigs we kept always did a very special line in baleful stares, ensuring that I felt deeply guilty delivering them to the sl— abattoir. To make matters worse, they were always treated like social pariahs by those in the next pen. Why? Because the Next Penners were clean, smooth, factory-reared examples of the genus porcine and

190

mine were muddy and hirsute and had lived alfresco. The factory jobs would look disdainfully down their extremely long noses and sneer as only a snooty pig can, as if to say, 'Are we by the slightest chance related?' To which ours retorted, 'Possibly, but at least we didn't eat any of our second cousins to achieve liveweight gain . . .'

Paul's misery and guilt were assuaged with the arrival of some spectacular chops and joints. Encouraged, and deserving of a night off amongst friends, he and Justine packed a few of Frank's remains in ice and loaded them into the hearse – a Renault estate – which travelled with unseemly haste, given the occasion, northwards up the M6. At Rowfoot, the pyre was lit and Frank, after a thoroughly decorous cremation, was sluiced into the afterlife, on which it seemed best not to ponder too much, with some fruity, kick-ass Ozzy Cab Sauv (mmm, I'm getting interesting aromas of Whiskas cat food and depleted uranium with a whiff of black pudding here). A pig could not have asked for a better, more appreciative send-off and naturally, as the wine flowed, so did the reminiscences.

'He loved Scotch pancakes, did Frank,' gloomed Paul mawkishly.

'Pass another sausage,' said Justine, smacking her lips.

Paul wistfully told of how he had constructed a special transport crate for collecting the weanlings and, of course, placed a goodly quantity of waterproof sheeting underneath – is that too much information, perhaps? The directions were crystal – 'When you think you're lost, you'll have arrived' – and pigs having a poor reputation for profitability they were absolutely certain they were lost when they found a swimming pool and an Aston Martin where the directions ran out. But they weren't and Paul, Justine and their new acquisitions travelled home, Paul placing a tentative order for a new pushbike (Aston Martins being for royalty, posh pig owners and flash gits) shortly after Frank 'n' friends had settled in.

Transporting pigs, we all agreed, was not entirely a straightforward job. It starts with impenetrable bureaucratic complexities and finishes with dung. So as the evening wore on we fell to telling pig tales, some taller than others and each slightly more alarming than its predecessor. The cleanest recalled a chap who had collected an outsize boar, lifted it (possibly this procedure involved a crane or at least a hoist) into his pick-up and set off for home. It was not a happy or a peaceful journey. With no inflight entertainment the boar became first restless, then agitated and finally furious, tearing at the headrest with a set of terrifying teeth. Our imperilled hero took evasive action, driving round a roundabout six times at speed until the boar was so dizzy it fell over. The rest of the journey was completed without incident although I suspect that he wove a fairly erratic path as he struggled simultaneously to preserve the integrity of his upholstery and to keep the boar from regaining his equilibrium.

Away from capsizing pigs, the original Big Frank had given his name but not his approval to Paul and Justine's rare Oxford Sandy and Blacks. 'You don't want them pigs, lad,' he advised Paul. 'They're no good, them pigs.'

Frank 'n' friends (not a Furter among them) were reared organically and did piggy things outside – grew spiky hair, got sunburn, enjoyed mudbaths and tried unsuccessfully to escape. Meanwhile, Paul and Justine established themselves as World Sausage Experts, forming firm opinions on texture, taste and size. Big Frank the biped still thought they were a mistake. 'Cash 'em in while you can. Get rid of 'em, lad.'

Weeks further on, Big Frank conceded that the pigs had done 'aw right', and now that he is the proud owner of a slab of belly pork he is a proselytising convert and will bore the pants off anyone who cocks half an ear, asserting that Oxford Sandy and Blacks really do make very fine pork.

As the mortally wounded sun bled into the horizon and the bats started to swirl and swoop, we talked of the future. For now, Paul and Justine have decided not to venture into the weird and wonderful world of pig showing, which involves a pig, predictably enough, and a human wearing a white coat and carrying a board with which to guide the swine. Hard as we tried, none of us could make any sense of it at all, but we did think that the bloke who wears the white coat ought to do up his buttons down the back. We couldn't explain the existence of the hand-held board either, though of course it could be an IKEA over-the-drainer chopping board for later use, and there seemed even less purpose to the shepherd's crook sans handle held in the spare fist, particularly as it's a pig being ushered about, not a sheep. It is funny how everything becomes impenetrable after a couple of glasses of vino collapso.

We did rather better with discussions on plans for a small breeding herd. A boar of their own, perhaps, ideally one without expensive tastes in designer upholstery, the better to preserve pick-up interiors?

'I expect so,' replied Paul, 'and I've made some enquiries about pig AI.' So did I, once, and when my eyes popped I stopped enquiring and left well alone. 'There's a bloke near us who does his own AI and he says that if anyone could see even the top half of what you were doing – or what it looked as if you might be doing – you could be locked up. There's another complication – everything comes through the post.'

There was a pause while we all digested this information along with a very good spare rib in barbecue sauce.

'The trouble is that when he gets a delivery, the postmistress shrieks at him, "Mr Potts, your semen's in . . ."'

You could see why it upset him, couldn't you?

Although I just do pigs by proxy now, having wimped out of keeping them years back, my enthusiasm for these engaging, intelligent, inquisitive creatures remains undimmed. Pigs deserve better than a dismal crated existence. We never kept Oxford Sandy and Blacks as Jus and Paul do; we kept Tamworths. Why? Mainly because I had an especial fondness for them as they are red-haired. Just as I like people to mirror my eccentricity, I like my animals to be auburn. It's not much to ask.

We used to keep our Tamworths in the small garth

behind the byre. They scratched merrily against the apple trees, salving their itches and dislodging tasty snacks from the branches above. I could watch them from the window as I packed up cream cheese and bottled cream in the dairy. We sold to passers-by at the farm gate too. They liked to come into the old dairy and marvel at the machinery, unchanged in principle for centuries: the churn and the butter worker, the separator and the settling jars. One visitor, although charmed by the bucolic view from the dairy window, had his own reservations.

'This Jersey cream,' he queried, glancing from his newly purchased pot of cream to me and beyond to the slumbering Tamworths in the garth, 'does it have the same effect on all of you?'

True, you wouldn't want to come across a Tamworth pig in a dark alley as they are probably closer to the wild boar than they are to the modern domestic pig, but I like them no less for that. But here's something you might like to ponder: pigs are omnivores. And you know what that means: they eat absolutely anything, making them ideal partners in crime if you are planning the perfect murder. Just think, you could shoot/strangle/hang your bitterest enemy upside down from a beam until they croaked and feed it/her/him, including crunchy bits, to pigs (though take care – if they were bitter and twisted individuals the swine could end up with nasty indigestion) secure in the knowledge that no tell-tale signs would remain to implicate you.

I was immensely reassured to see that no less a luminary than hot-to-trot blonde bimbette Lily Savage's alter ego Paul O'Grady has committed to taking exactly this course of action following the theft of his close pal Cilla Black's spectacular collection of rocks. At the launch of Cilla's autobiography Paul was quoted as saying: 'I've just bought myself a gun. After what happened to Cilla, I'm not taking any chances. If I'm lying in bed and any gob-shite burglars are in my house thinking I'm not going to do anything, then they'll be in for a shock. I'll shoot them in the kneecaps and feed them to my pigs. I'm with Tony Martin on this one. If you're in my house and you shouldn't be then I'll shoot you, simple as that.' Quite.

I rather like that. I like its unequivocal inelegance and its commitment to crime prevention, and it's nice that, just for once, someone is telling it like it is rather than pussyfooting about in the name of political correctness. I can even forgive Mr O'Grady for having nearly propelled me into a monumental shunt on the M6 southbound as he told Radio 2 listeners how he had mixed up his vitamin pills and his new pup's Bob Martin's worm tablets. If he fills in for Jonathan Woss again, though, I hope he will be just a little more thoughtful about the RTA-inducing properties of his commentary. In the public interest and all that.

June

All the stock are out now and enjoying the summer grass. And as we all fret about energy and its generation, I wonder why no-one has thought of harnessing cow-power, or at least the utilisation of the methane they emit when they burp. And given the frequency of cattle-burps, this is a real untapped resource of environmentally sustainable fuel. Everyone is exercised about global warming, but cattle-belching – not a lot said about it, is there? Re-examine proper coppicing, and harness belching bovines to the National Grid, and we could forget about fossil fuels and nuclear energy. Failing that, build a waterwheel. But check that you live by a watercourse before you reach for the axe, won't you?

I SWUNG TO MY heart's content. That is what you do – what you have time to do – once lambing is over, you've spot-sprayed the nettles in the long meadow, not daubed the gates with creosote because it's illegal now, Mrs Mopped the holiday cottage for the incoming guests, fed the pet lamb and evicted TV Tess from the greenhouse where Malcolm's tomatoes are growing – burgeoning – and she is not supposed to be.

How did I swing? On my garden bench, bought in a frenzy of profligacy, an attempt to liberate myself (and my sheep) from the petty bourgeois affectation of having a bank account displaying black figures. I have always wanted a garden swing, ever since I was a child and Auntie Rose had one; I used to ask to be allowed to swing in it even in the winter and when it rained, so fixated was I with its gently soothing rhythmic motion. And now I have my own. I shall not be inviting anyone else to use it even though it says it seats two.

This afternoon, I settled down with a cup of tea and girded my loins for some serious relaxation. I saw one blue tit mistake its reflection for an adversary and nearly perform a frontal lobotomy on itself, a blackbird pause from perfecting his penny whistle impression to collect first a worm and then a shard of last night's leftover ciabatta (half price, Tesco, no olives), five sparrows indulge in a bit of communal bathing rather as humans do in Budapest, a chaffinch fall off the feeder and Blossom play dead so convincingly I ended up walking up the field to see if she was still alive. I had to prod her to wake her up: pretty relaxed then, Blossom.

I pause on top of the hill and gaze across the valley. It is a glass-lipped vision of loveliness, the trees sharply delineated against a sky of cerulean purity. Grassy splays and folds spread out like the folds of a great silken skirt, edged with hedgy darkness and sliced through here and there by shadows and shafts of sunlight.

I rather like this time of year.

*   *   *

And you know what they say about procrastination? Never put off till tomorrow what you can do today because if you like doing it today you can do it again tomorrow. And that's what we've been doing today – exactly what we did yesterday. Sitting in the garden, basking in the welcome sunshine.

We never go far on bank holidays but today we had planned a foray as far as Greystoke, ten miles distant.

'Do you want to go, or shall we just . . . not bother?'

'Don't let's bother.'

'Fair enoughski.'

So there we sat, drinking tea, having lunch, dozing slightly, reading the paper until evening.

Evening was set to be busy, you see.

And as usual, I have panicked unnecessarily. Well, why worry when you can have a real, full-scale, bells and whistles panic, that's what I want to know.

I was bothered that Blossom, having not been in harness since last September, might have forgotten what it was all about. Although she had been driven through the middle of Chester-le-Street and we had taken her out ourselves, I was concerned that she might have forgotten it all and recall instead some of her early aversion to pulling a trap. Blossom had been a 'right sod' to break and that's quoting one of the mild-mannered souls at the RSPCA equine centre, so I think we can take it that it wasn't easy. Possibly she misunderstood the term altogether; perhaps she thought she was the one who had to do the breaking as she had demolished a cart and a set of harness in the process, then stood up on her hind legs in a quite unnecessary display of exhibitionism and waved at an alarmed gallery. I didn't see myself coping with any of that, so it had been arranged that she would be sent to boarding school at Gilsland with Charlie Parker for a week of driving lessons. And have a red letter L stuck on her bum, too.

In the event, Charlie had spent so much time re-roofing the barns in the courtyard at Gilsland in preparation for his new venture, a Working Dales Pony Centre, that Blossom's revision course had been delayed. Then, wondered Charlie, did she really need it? How about if he tripped down to Rowfoot and we shackled her to the exercise trap – at least if she gave that a good kicking it wouldn't involve remortgaging the house to repair it, as it surely would if she booted the Bennington – and asked her what she thought? I was persuaded and Charlie fetched up on the evening of bank holiday Monday, looking cheery, weathered and ready for anything. The harness that had fitted Blossom in September needed a little readjustment, our cob having, well, Blossomed in the intervening months.

She looked at me cheekily from behind the blinkers as much as to say, 'Oh, today's driving, is it?' And she stood like a little rock as the harness was fitted to the natty little blue trap. Then she pulled it, Charlie and Malcolm out of the drive and away up the hill into the village. She walked, she trotted and they all came home.

Charlie said that she was 'sound as a pound', 'a grand little lass'. Such words from Charlie, a man not greatly given to effusiveness, was praise indeed.

And I am hanging my head in shame at having mistrusted Blossom so unfairly.

Had she needed a short revision course, though, Charlie was undoubtedly our man. He lives up in England's Big Country: vast billowing skies, majestic sweeps of field and fell, ancient stretches of Hadrian's wall, die-straight roads unaltered since Roman times bar the occasional lick of tarmac. The landscape around Charlie and his wife Gina's home at Clarks Hill near Gilsland probably looks little different now from when the farmhouse was first built, so it is fitting that it is against this timeless backdrop that they are reviving old skills with traditional breeds of livestock.

Charlie, who had worked for years in the forestry business, recognised some little while back that the times they were a-changin': traditional skills were being replaced by bigger and bigger machinery. Realising that he had to invest massively or shift gears altogether, predictably enough Charlie the horseman and craftsman forester rejected the mechanical option and changed direction, working his Dales ponies on specialist projects in terrain where machinery could not reach: kinder to the forest, kinder to the planet. Then Charlie's knees began sending out distress signals. 'This time last year, I had been in and out of hospital four or five times,' he told me. So once again it was time to set a subtly different course on the compass.

The mellow stone buildings at Clarks Hill have been restored and repaired, the land is in stewardship and an extensive programme of hedging and ditching is under way. Water comes from a pump house supplied by a spring, but the surrounding land was suffocating under a lavish carpet of dockings and weeds. Spraying was out of the question as it would contaminate the drinking supply, so another answer to this small crisis was essential: an environmentally friendly one. Enter Bob and Elsie. Bob and Elsie are pigs; rather unusual pigs whose presence at Clarks Hill owes much to a chance meeting between Charlie and a fellow at a forestry demonstration up at Lockerbie who said he had just the pigs for the job.

Well, he would say that, wouldn't he?

This pair of Tamworth cross Wild Boar swine are about the nearest twenty-first-century approximation to an Iron Age pig. They will scarify the land. I can vouch for this: I met Bob and Elsie and they scarified me. A matching pair of auburn porcupines with Number Three haircuts, Bob and Elsie are sure to prove accomplished weed-management contractors. It's what pigs do best. The bit I'm still unsure about is how they'll avoid contaminating the water supply, but I guess that some things are best not delved into too deeply.

With Tamworth pigs classified by the Rare Breeds Survival Trust as 'endangered' and Dales ponies as 'vulnerable' it followed naturally enough that the cattle at Clarks Hill, belonging to the Parkers' son Roger, should be traditional Belted Galloways and Dexters. There are a few goats and a donkey too but it is the Dales ponies that are at the hub of plans for the future.

During their lengthy spell in Yorkshire and a brief period up in the Borders, Charlie and Gina established a successful Dales pony stud concentrating on roan-coloured (pinky grey or bluey grey) Dales ponies rather than the more usual brown, bay and grey. They have also experimented successfully with crossing their Dales ponies with hefty coloured cobs; both bloodlines will also provide the workhorses for the Working Dales Pony Centre. The Centre has evolved from merging all the things Dales ponies do so well: commercial timber contracting, 'snigging' timber and harrowing in chains, as well as conventional driving to a trap. In response to a growing interest in the revival of these ancient skills Charlie has run some open days demonstrating them, some independent courses and some under the auspices of the Forestry Contractors' Association, passing on nuggets of his own experience to new enthusiasts. And once he's finished dancing about on all the roofs at Clarks Hill, and converting dead barns into cottages, Charlie will have more time to teach both day and residential pupils.

Then there's his taxi service. You know that sign on the back of all black cabs saying something about Hackney Carriage for hire? Well, the truth is that most city councils haven't issued bona fide Hackney Carriage licences for nearly half a century; what they have issued is a great many licences for diesel-powered nothing-to-do-with-Hackneys-at-all taxis. Charlie, of course, has the real thing in mind and getting to grips with this particular reality is proving quite a challenge for the local city council. But fear not; they say it is all in their capable hands.

The plan is to operate a horse-drawn carriage along

the Roman wall; more Dales pony than Hackney horse but a carriage none the less. Routes and termini might include Birdoswald, Gilsland, Banks Turret and Clarks Hill itself, where a visitor centre will display harness, implements, carriages, historical documents and by-gones.

There's an unmistakable nostalgic whiff about Charlie's latest acquisition, a glamorous dark blue victoria, too. It's the kind of carriage fairy princesses use for everyday flying about and spellbinding but Charlie drives brides to weddings in it. His imperturbable driving cobs Gilly and Jack bring that bit of sanity and calm normally conspicuously absent at weddings to the proceedings. Charlie is none too sure about 'doing' weddings. 'Can you see me in a top hat and tails?' he asks, slightly unsure. Actually, yes, Charlie. And with just a little facial adjustment, you'd be brilliant at funerals too.

If the ponies and pigs at Clarks Hill are all in gainful employment, the dogs are definitely not. Instead of a Dalmatian to accompany the victoria or a fogeyish Labrador to work on the moors, there is one rough-haired strawberry blond job with the kind of lower jaw that would make a canine cosmetic dentist very rich indeed and Mabel, a wire-haired (very nearly as wire-haired as Bob and Elsie, but not quite) miniature Dachshund. Mabel has no visible means of propulsion, her recessed legs only becoming apparent when she rolls on to her back and sticks them up in the air, in confident expectation of having her tummy rubbed. Charlie affects faux disgust: 'I told Gina she couldn't have one of these things living in the country. All last summer I told everybody, "It's not mine, it's wife's."' He bends to scratch her anyway. I think he rather likes her really.

'If all else fails,' says Charlie cheerfully, 'we'll breed a few more of these and stage Dachshund racing.' Somehow, I can't see that happening – the failing, that is, not the Dachshund racing.

And whilst Blossom didn't need to go to Charlie's reform school, we sure needed his expertise at Rowfoot, however briefly.

There's a kind of madness abroad. But instead of staying abroad where it belongs, it's creeping ever closer to home. And I try to be tolerant, I do really; after all, it takes all sorts to make a world. But then, as the one-time sub-postmaster in the next village once said to me, very wearily: 'Why do we get them all in Armathwaite?' He had a point, I guess. Therein lies the rub: they all *want* to be in Armathwaite, or Ainstable or some other pretty village, miles from disputes fuelled by mad neighbours' rampant leylandii hedges.

They've read the book, seen the film or at least watched repeats of *The Good Life* and subscribed to the magazines – you know the ones, where we all wear gingham frocks while gathering rosebuds from organic bushes and eggs from free-range hens and do card quilling in our spare time.

Methinks it is time for a reality check.

Many a delightfully situated ye-olde-stoney-cottage with roses clambering up its walls and a bucolic view from its windows – cattle grazing peaceably in the field at the bottom of the garden – turns out to be rather different in reality. The roses may dissuade burglars clad in thin chinos but the cows get up in the middle of the night – well, 6.30 a.m., anyway – for milking. And milking itself is a machinery led and driven process, tractors scraping out slurry from between the kennels, with vacuum pumps, coolers and tankers all clanking as noisily as anything on an urban industrial estate. Out in the fields it is no quieter nor any more peaceful. The corn is no longer hand-flung by lovelorn gals in gingham but drilled by an assortment of fantastical machinery. Grain silos rattle and roll. Noisily. It might not have been quite what potential escapees had in mind but it hasn't put too many off. Even so, several factors have conspired to change the complexion of the average Eden village beyond all recognition over the past twenty years or so.

Ainstable may look like a time-warped rural idyll and mostly, barring the odd contemporary horror and the occasional neo-brutalist bungalow squashed on to a site the size of a bathmat, it is. But scratch the surface of rolling fields and lowing cattle and you'll find quite sane people running businesses from home via t'internet and bathing in solar-powered hot water.

Undeterred, fed up with commuting and anxious to get in touch with their inner peasant, they flee the suburbs, gleefully swapping exhaust fumes for the delicate aroma of spreading slurry and wailing sirens for gas-gun bird scarers.

So, here's my very own Brief but Essential Guide to

Differences 'Twixt Town and Country: not the snappiest title, I grant you, but at least this little treatise doesn't mess with the truth in the way the glossy mags do.

Food and drink. You can get takeaways in town – Indian and pizza for the faint-hearted, sushi, Thai, Vietnamese and railway sandwiches for the brave – delivered any time, day or night. In the country you can't. It's that simple. So if you like sophisticated surroundings or restaurants with flock wallpaper stay in Surbiton.

Facilities, services and many other things you do not need. In town, your flat or apartment will have all mod cons, including mains gas, proper sewerage systems and electricity that will rarely fail for longer than a whole episode of *Coronation Street.* There will be pavements and street lights. Litter will be picked up by a little man driving something yellow, battery powered and mounted on fraying barrow wheels with a brace of partially squashed hedgehogs pasted to its under-carriage. You will often see policemen and notice that they are scarcely old enough to be allowed out to play on their own. For all this – the mod cons, mains gas and sewerage, the pavements, street lights, litter collections and police presence you will pay council tax. Natch.

In the countryside, you will have no mains gas even though the pipeline goes across the hill at the top of the village ('just not cost-effective to run it down to the village, madam') and power cuts will invariably occur at Christmas for approximately four days. God help you if your house was designed by an architect with scant understanding of the vagaries of rural electricity supplies, a topic upon which, no matter how many times it is demonstrated, modern developers suffer an inexplicable amnesia. Finding the cause of power failure, though, will provide hours of endless fun and diversion.

It is now that the wheezing monolith previously known as Aga, Ray Burn or Stanley – which you regularly threaten to abandon on the municipal tip – comes into its own. You may never previously have considered working in a soup kitchen but now you will find yourself at the hub of one as neighbours from the Amnesiac Architect's new-builds descend en masse, grateful for soup, hot tea and somewhere to nourish incipient chilblains. If it is Christmas, the turkey will fly magically from one oven to another like a gilded vagrant, the puddings will simmer happily, blip-blopping on the top of the stove, and the dogs will park themselves immovably on the rug in front of the cooker, secure in the knowledge that you will make strenuous efforts not to stand on their paws or tail as you drain, sieve, test for doneness, make gravy and generally footle about with the dinner.

Your sewerage arrangements are best not in-vestigated unless they demand your attention – and you will *know* when this occurs, trust me. When your drains block, you must lift your own manhole cover and stick your nose into the abyss. Use a peg – you know it makes sense. Now shovel the suppurating swill into the waiting ranks of buckets and stick a long rod – or several long rods if it is a serious problem – down to free whatever it is that is stuck in the pipe. As far as the solids in this fetid mass are concerned, do not, under any circumstances, venture into the delicate Blame Territory of 'this one's yours', 'that's one of mine' and do not, whatever you do, ask your neighbours (if you have any) to send one down with a flag on it to monitor progress. It is not done for peasants to weaken like this. Be stoical at all times. And keep rodding . . .

There are no pavements and few lights, the litter collection is a 'community initiative' in which you are expected to participate and you will rarely see a police-man. And when you do, it'll be some bloke in fancy dress.

You will share this much with your city friends, though: you too will pay council tax. Lots of it.

Builders. In town, a builder is about as reliable as Mystic Meg's predictions. Perhaps I am just chary because her last for Sagittarians warned that a boulevardier named Esme would invite me out to dinner on Tuesday when I would be too busy doing a Life Laundry on the mixed manmade cycle to accept. Lucky shoe, she said, was moc croc stiletto. I ask you? Laundry on a Tuesday . . . it would just never happen.

Your townie builder turns up when he feels like it, arriving in a van with number plates that are obscured by generations of dirt, indecipherable or non-existent, does work that never needed doing in the first place, overcharges for what he does and leaves without saying whether or when he might return.

He is unlikely to be called Bob.

In the countryside, your builder will be called some-thing sensible like David or Eric or Ted. He has hands

209

that, when spread, are the size of an elephant's foot, and is conversant with the difference between hard and soft woods and the fact that MDF differs from both in quality and price. Tutored in the same tradition as our Eric, he too may have served his lengthy apprenticeship fashioning coffins for the dead wealthy, or come to that the wealthy dead, of the area. You can give him keys to your house without wondering whether he will open your mail, ravish your daughter (even that might not be such a bad thing: my mother always advised me to marry a man who could paper and decorate), or kick your cat. You will not, if you are as fortunate as we have been, even care if he remembers to give the keys back at some time within the next two decades. He works long hours, with skill and care, and bills you with honesty. The only tricksy bit is persuading him to actually prepare, send and receipt the wretched bills.

The blacksmith. A town or city blacksmith makes pot-bellied balconies, fenders for baronial fires (which is a shame if you live in a Barbican loft conversion), and dinky door furniture. In the vanguard of the British Folk Art movement he refers to himself, without a hint of irony, as an F-Artist or artisan, and although he doesn't tell you this bit he is quite possibly a charlatan too. Out in the sticks, a blacksmith is the bloke who shoes horses. Your obligations are as follows: never expect him to materialise at short notice except in dire emergency (horse trial the next day, sandcrack on the near fore, lameness or a lost shoe; and if the last happens he may refuse payment, in which case make

210

sure you can pay him in either lamb chops or freshly baked bread) and always, always pay him in used notes except on the lamb-chop-currency occasions. He is still perfectly capable of rustling up pot-bellied balconies, fenders for baronial fires and ditsy door furniture – he just can't see the point.

The wake-up call. In town you don't need an alarm clock as the traffic stacking up outside will wake you at seven twenty precisely. It's tough if you need to get up earlier than seven twenty but you can rely on the dust cart on Tuesdays, the road sweeper on Wednesdays and the excitable burglar alarm at number 79 for most other days of the week so you need very little inventiveness where excuses are concerned, though anything involving aliens, organised crime and carbon-dating is always worth trying. In the countryside, you may find that the famed silence is actually a myth, what with stampeding horses whipped into a frenzy by marauding hot air balloons, tractors towing empty silage trailers, angle grinders, mig welders and quad bikes firing up at dawn, not to mention crows, collared doves, just-weaned calves hollering for milky breakfasts and lambs temporarily estranged from their mothers. But don't panic just yet. Buy some earplugs, turn up the radio and you'll learn to live with all of it . . . if you stay long enough.

Public transport. In town, you are spoilt for choice. Will it be bus or rail? Or shall we just fly as the nearest airport is within an hour's drive from home? You may even be able to burrow down like a metropolitan fox and use the Underground; taxis, while financially ruinous, are at least *there*. You neither need nor want a car but you have one anyway, and a resident's permit for the privilege of seeing it parked outside your front door seven nights of the week. It may, though, be there permanently not because of its advantaged status but because it is on blocks having been relieved of its wheels by a gang of youths.

In the countryside, there is only one question about public transport? What is it?

If you have a *pet* in town, it is either something exotic like a gecko, pointless like a pet rock (what was that all about?), or demanding and troublesome like a dog. Walking the dog is nearly as demanding as the school run as 'parks' are mostly dog-free zones or at the very least dog-poo-free zones, meaning that, no matter how grandiose, elevated and senior you are at your place of employment, no matter how much you bark 'Get me the Queen' into your state-of-the-art Nokia phone-with-camera-with-instore-cappuccino-maker, to your dog you are just a personal pooper-scooper.

The dog will spend hours staring out at traffic and envying stray and bedraggled cats their freedom. When you and Fang (a Pomeranian, a fashionably furry accessory) go for a walk, he will enjoy rooting amongst the discarded hypodermic needles and styrofoam cartons containing the mummified remains of burgers that taste like factory waste and fish fingers from piscine multiple-amputees who thoughtfully parted with their digits just so that he could be bilious. He does, though, meet a satisfying number of Shih-tzu girlies and fluffy Yorkies with ribbons in their hair, even if he knows that Yorkies are ratters at heart no matter how many embellishments they stick in their locks.

Country dogs are Labradors, or collies, or Jack Russells that go ratting properly sans frippery because frippery just gets in the way. Springer spaniels abound and are all completely demented. Irish setters are mad, but don't do as much bounding. Pointers point, lurchers lurch, terriers terrorise, a Dachshund is one dog one and a half dogs long and Dalmatians run between carriage wheels, risking turning their paws into pâté. And dogs that hunt are hounds, not dogs at all. So now you know.

Horses are completely off limits in town, the cost of livery being astronomical, but just about manageable in the country. You might go for an Arab with go-faster stripes or a stylishly splashed coloured cob with such extravagant feathers that the blacksmith might suggest that you cut the feet and knicker bits out of your tights and pull on the leggings bits so that he can get his task completed without all the plumed impedimenta getting in the way.

You don't *work from home* in town because that would be silly when you can get the Tube, and anyway you don't want to because you would miss out on office politics. You would have absolutely no idea who is doing what with whom and how often. In the country, you work from home because it saves on office rental, you can sneak a lie-in without anyone knowing and you do all your work via email. The flip side is that you are permanently available.

Finally, the *indigenous population* in the countryside live in satellite encampments never far from the tribal base, often in captive breeding projects. These isolated habitats provide safe havens from progress, technology and contemporary mores – all of which are energetically eschewed – and ensure certain preservation from extinction. And, in lifestyle terms, the wildlife are nearly the same . . .

But once settled, made over and gloriously self-sufficient in eggs, vegetables, wool, meat, milk and muck, you'll have masses and masses of free time. What is more, your brains won't be fried by mobile phone waves. We do have mobiles out here but they don't always work. In Ainstable, we genuflect in quasi-religious obeisance to the mast on the hill and hope for the best. The best does not always happen. So, without the constant intrusion of mobile ring tones and absolutely no need to yell 'I'm on the train', what do we do? You may well ask. I spend a great deal of time capturing pet lambs, swinging on a garden seat and playing on t'internet myself.

If you make the Big Move, you'll be asked as often as I am, 'Do you miss London, then?' In truth, there is much to be said for London and now that I go back as a tourist, not expecting the city to sing out to me, I love it. But live there again? That is a very, very different matter.

The very idea fills me with horror. It is the stuff of nightmares. Real nightmares, that have you waking up with your fringe stuck to your forehead with sweat, whimpering at visions of animal welfare charities bearing your creatures – without you – off to a place of safety.

It all started with the Ten Year Dream.

I dreamed that the government of the day issued a directive that everyone had to go back to live where they lived ten years previously. No appeal was permitted, no argument brooked, no special circumstances considered, they said: just do it.

So, we loaded three horses, twenty-five sheep, six cows, two stirks, four pigs, a couple of dogs, five cats and ourselves into our lorry. It must have been a bit of a squeeze in the back there, because it was originally a GPO van converted to transport two full-size equines, but that's one of the funny things about dreams and nightmares: the details never seem important.

We arrived back in Walton-on-Thames and unloaded the livestock. Stranded on the pavement outside our three-bedroomed semi, I worried about whether to take them through the house – risking carpet damage, obviously – or the utility room to the garden.

I know, I know: it's completely bonkers. But in the loneliness of the witching hours, it felt so very real.

After ten years of living at Rowfoot had elapsed without a government directive making unreasonable demands about dwelling places, I began sleeping a little more soundly. Twenty-two years on, I still occasionally have the dream, but its intensity is less bothersome than it used to be.

This hot, dry spell has lasted so long that farmers are already crying out for rain. It is, though, great for holidaymakers in the cottage. I've been telling them that it is always like this, hot, sunny and dry, and that it only ever rains at night in Eden. My nose goes back to its usual length after an hour or so, usually.

Sunday, and the weather is still wonderful. So we take the papers into the garden and settle down for a read. There's something in the paper about a collie with a vocabulary of over two hundred words. He'll go and pick up a banana or a dinosaur or a werewolf. (People surround themselves with the oddest things. I have a banana or two knocking about but I'm fresh out of dinosaurs and werewolves this week.) My friend Jane contends that this collie's vocabulary is substantially more developed than that of the average man. Indeed, Jane reckons that there are very limited windows in a man's life when he communicates in anything other than grunts.

'Until they're six or seven, they just grunt,' she avers. 'And they grunt right through their teens. Maybe they talk between the ages of say, eighteen and twenty-five, but only as foreplay, and then once they get into their thirties they revert to the grunting again. They've had lots of grunting practice by then.'

So Rico the collie is remarkable: particularly as he's a boy collie dog. The German team monitoring the doggy prodigy warn against rushing out and buying a collie dog; they don't say anything about the warning's being particularly targeted at garrulous women but I bet that's what they are thinking. 'They are working dogs. They are high maintenance, professional dogs that need four or five hours' attention a day.'

I read this out loud but Gyp and Tess miss it because

they are both sound asleep. Funnily enough, though, Rico looks a bit like Tess.

Paul and Justine arrive with as fine a chunk of pig as anyone could wish for.

'I'm really proud of this pig,' says Paul happily. 'He's Big Barbara's son and he's called Bloody Lucky. His mate Lucky has gone too and then of course there's F— Lucky – he's still at home because he escaped from the corral so he has a couple more weeks left.'

We have a bit of catching up to do as they've moved house, bought more pigs, lost a cat (an injured stray who gratefully adopted them after they'd stumped up for his vet bills and then, inexplicably, decided to end it all in the road outside their new house) and started doing farmers' markets ('We sold out in two hours at our first one' – no surprises there).

Large Black pigs have emerged as their favourites in terms of quality and tractability so it's those they intend to concentrate on in the future. 'We had some Gloucesters but they were moody little buggers,' says Paul. 'And Middle Whites are ugly.'

I want to know how they are getting along with that pig AI we talked about on their last visit. It is a notoriously precarious practice.

'Not great really. The twice I've phoned up and ordered semen, the boar hasn't felt like obliging, so we're using whatever young boars we happen to have at home. But in the long term we need AI for the pedigree breeding.'

'Does someone come out and do it, like with cattle,'

asks Malcolm, 'or do you have to have special training?'

Paul says no training is essential, but 'I asked this old farmer and he said, "Well, you're a fella so you'll have some idea of what is supposed to happen."' Justine raises her eyebrows but wisely says nothing. '"And since the sow's a woman, if you don't get it reet, she'll let you know . . ."'

And that's it, the Short Course in Pig AI Techniques.

It's a new skill to add to his portfolio. Amongst the many others Paul picked up when he was a stud manager was fencing and he tells us he'll be doing some of that too, for reasons not entirely unrelated to another aspect of his earlier employment.

A chance purchase of the *Racing Post* in Derby Week and the name of one of his former charges leapt out at him.

'He was a cracking little colt, beautifully bred, but backward and he needed time. Lots of it. And at a hundred and twenty pounds a week, any trainer will tell you to give 'em a bit more time.' Paul smiles ruefully. 'So I said to Jus, "I think I'll have a bit of a punt on him."' And, predictably, Justine went up the wall, across the ceiling and down the other side. She pointed out – sensibly because she is Woman – that they had just moved house and possessed a vast number of pigs to feed and that doing something as reckless as gambling – gambling, for goodness' sake – could not possibly be a rational way to spend their hard-earned cash. And Paul, good and faithful husband that he is, listened and understood and, of course, agreed with every word.

And then he went and placed the bet.

Having a gambling habit every bit as reckless and feckless as my own (I once put a whole pound on a Placepot at Kelso), he backed the colt at two pounds each way. And it trotted up at 50–1.

'He said I could have thirty pounds of it,' spluttered Justine.

Paul thought this was reasonable enough. 'I've got enough for ninety fence posts.'

Hence the fencing.

Paul is a committed pig farmer these days. Indeed, he has only wavered once in his evangelism. Why? Well, if you find yourself in hospital, your future health reliant on the expertise of a delightful Muslim specialist and with a Mr Goldblum in the next bed, it is probably best to keep quiet about being a pig farmer.

Animals. Healthy bank account. Mutually incompatible objectives, rather like wanting to take a luxury cruise and circumnavigate Switzerland. Nice ideas, but never likely to happen contemporaneously.

Let me explain: stay with me, if only out of morbid curiosity.

Gyp had been quiet. Nothing remarkable about that, as she is always quieter than Tess, but this time she was quieter than usual. I put it down to a combination of age and the sweltering heat but then she began to – how can I put this politely? – smell. Worse than a skunk's oxter – a real honky pong.

So bad that something, surely, was wrong. So we carted both of them off to the vet. The vet took a look and made a swift diagnosis. 'Pyometra, and we need to get it sorted quickly.' Gyp was already toxic. Poor little dog.

'Still,' said Ann, the vet, 'no waiting on a trolley for you, dear.' The drip was in before we left for home. You don't get that on the NHS.

Ann operated within the hour and whipped out Gyp's hideously infected uterus. She was on a drip overnight, and although she had come through the anaesthetic we were not greatly encouraged by the response to our phone call next morning. 'She's not out of the woods yet.'

Committed to going to Workington and feeling that sitting around all day waiting for the phone to ring was less than a good idea, we elected to stick to the plan and dosed Tess with Stugeron to make her travel-fit. I don't think that metropolitan Workington was much to Tess's taste, though it might just have been the shock of being in a town for the first time in her thirteen years. She probably found it all a bit alarming – an unending procession of people's ankles and lower legs must be a pretty shocking panorama for even the most well-adjusted dog. She responded to it by pooing on the first three street corners we came to, necessitating emergency use of the scoopie and, when I ran out of plastic bags, the *Daily Telegraph* financial supplement.

In the Derwent Bookshop Tess relaxed, meeting and greeting her public with aplomb and a very waggy tail indeed. She persuaded one browser to purchase a copy of *The Funny Farm* but seemed slightly surprised and not a little disappointed when her offer to sign it with a

pawprint was politely declined. No matter. She trotted
out happily and we progressed to Whitehaven and the
chippy – very fine chips, and a superb slab of fish
the thickness of an average lorry tyre – then slaked our
collective thirsts at Zest Harbourside, the bistro arm of
the ultra trendy Zest, where Tony and Cherie and at least
one literary peer of the realm dine, but we didn't let any
of that put us off.

If you have a dog, you *need* to go to Zest Harbourside
because, unusually in twenty-first-century Britain, dogs
are positively welcomed there. We had beer and cider;
Tess had a shiny stainless steel bowl of water – had she
requested ice and lemon I have no doubt that it would
have been delivered – and a wrap of dog biscuits. What
a very fine policy.

On the way home we visited Gyp, kennelled at
the vet's, but she looked dreadful. Next morning she
seemed to be faring better physically but not mentally; it
is a funny thing, that we manage to forget the emotional
life of a dog. Gyp, normally in a position to choose
how much company she has, where she sits and when
she sleeps, suddenly had none of that freedom. Caged
to recuperate, she seemed to find the effort needed to
recover beyond her and looked so listless that she
appeared almost lifeless. Time, the vet said, to take her
home and see if being with Tess and in her usual
surroundings might act as the much needed catalyst to
recovery.

It did.

Although we had to lift her out of the car, using the
blanket as a sling to haul her up to the house, within

minutes she was looking about and staggering, slightly dozily, into the garden.

The following week, not wishing to be left out, Tess had four mammary tumours removed and Malcolm suggested that we should just leave the credit card at the vet's. Think of all those points accumulating on my Tesco Mastercard, I said brightly. He gave me one of those looks that said if I were any more stupid I'd need to be watered twice a week.

But the bright side is there all right: two dogs, in spite of being nearly thirteen years apiece, geriatric in canine terms and ninety-one in human ones, are positively bouncing. The biggest problem is trying to stop them doing anything daft. Sometimes you win and sometimes you don't: on Sunday I didn't. Gyp decided that she was sufficiently restored to leap the rails at the top of the flight of steps down to Armathwaite bridge. She got stuck. She looked round at me, as pained an expression on her doggy face as imaginable and her thought balloon reading perfectly clearly, 'Help. Quickly.' I extricated her, and she gave a shake and trotted off. I think humans make far too much fuss, myself.

In the middle of all this, the computer crashed. Or at least, the modem failed to dial up. It claimed it couldn't find a dialling tone so I rang my pal Bill. Bill knows about computers; I thought that he must have worked 'in computers' – necessitating a diminution to Lilliputian proportions, obviously – but he hadn't. Bill has a self-inflicted and intimate working knowledge of computers as a consequence of once having had an Olivetti with an inadequate guidebook. Instead of fulminating against the paucity of information, he took it to bits, played with its viscerals and put it all back together again. Bill is not normal, clearly, but he's a good mate to have. So he tinkered with it the way spotty adolescents used to tinker with bikes and sent me off to collect a new modem from Garfield in Penrith. Garfield is a man of few words but encyclopaedic knowledge and sorted out the modem, pronouncing the computer fit for normal duties.

'Working all right, 40Kbps,' Garfield said.

'19K at home,' I said, echoing Garf's verbal economy.

'You're DACsed,' he said.

There was a kind of finality about this. It wasn't the first time I had been told, but for years BT had refused to confirm, deny, acknowledge or, more important, do anything about it. I wasn't even sure what it meant, beyond the fact that it made everything very slow, but now that I had it confirmed by an expert I determined to do something about it. As soon as I had checked out eBay anyway . . .

Guess what? When the computer got home, and Malcolm had writhed about on the floor wrestling with first the computer wiring and then, after a very short time, the consequences of a spine fractured in 1981, it failed to find a dialling tone. Still.

We pulled out all the other extensions in the house, disconnected the computer from the DIY installation upstairs and reconnected it to the BT installed extension hole in the kitchen. Now, having your computer situated above the dishwasher is not, let me assure you, ideal, not least because of the ever-present threat of treading in the dogs' water bowl. You can't sit down either and that could become wearisome after an hour or so. If it worked, of course.

Then it didn't. And then it did, a couple of times. So I logged on to Wanadoo and signed up for broadband – and another new modem, of course. I would shortly have an impressive collection of these things – one that didn't work, one that worked intermittently and one that wasn't operational yet. Heigh ho, the wonders of modern science.

The computer sighed and died and we took it back to Garfield. It dialled up lustily. Garfield sent me home.

It didn't dial up at home, though; not upstairs, downstairs or in my lady's chamber.

Bloody thing.

Garf said he still thought the DACsing and BT were the culprits. Time to do battle. I phoned BT and asked them, in a friendly voice, to confirm that I was DACsed. Yes, that's right, they said. But it doesn't make any difference at all, madam, and even if it did these lines are only guaranteed for voice calls. So it's an unfathomable mystery why BT are marketing themselves as a company that provides an enviable internet service, isn't it? A bloke called John in Complaints said that was bunk anyway, and they guaranteed a minimum internet connection speed of 28K. Later on, much later on, someone else in the corporation denied that flatly, so I have no idea what is right.

I fulminated but I couldn't unpick BT's innards

because, wisely, no victim was immediately available. You really don't want to know the ins and outs of the next fortnight – yes, that's right, fortnight – but in the end BT arranged to come and unDACs us on Thursday 1 July. A man came to do it on Tuesday 29 June and said that my problem must be my computer, my software or my modem because it wasn't BT's line. Within minutes, though, we had another fault on the line and by Wednesday we had no dialling tone on the phone and, obviously, no computer dial-up either. Another engineer came and fiddled with yet more wiring and installed a different plastic box and lo! – the computer dialled up and the phone worked.

Have you lost interest yet? Why ever not? I did.

Anyway, it all made us realise how much we had come to rely on the large plastic thing with lots of wires – for banking, for shopping, for booking holidays, for playing about . . . As I don't cope with anything more technical than a shovel usually, I can only assume that I am now hopelessly addicted to my computer.

July

It's now that the local show season really gets into full swing. Some of these are huge affairs, with marquees, entry fees and corporate hospitality – a free cup of tea and a Hobnob in the bank tent in a poor year and a large Scotch in a good one – and others are small, local affairs. The competition is equally hot at both, whether for the Children's Monster Made from Seasonal Vegetables or for the Best Beef Bull. Never, ever pass up the chance to visit an agricultural show in Cumbria: as Life Experiences go, it is one of the most fun things you can do legally and fully clothed.

I T'S JULY. IT'S raining. Tim has lost at Wimbledon, England has lost at the footie and the cricket is on a knife edge. Are we good at anything any more? Tiddlywinks? Moaning?

Still, the computer is still working and the dogs are still alive, so things are, as some misguided politician once said, getting better.

*   *   *

It was high summer back in March and the sheep were sweating. Now that it's clipping time, the weather seers are predicting a bout of 'unseasonal' weather. Mr Fish even made a rather thin joke last night about a phone call he had had from 'a lady in Wales' – an oblique reference to his disastrous dismissal of the coming hurricane in 1987. All this, my friends, leads me to believe that it will be several degrees sub zero and that we will be buffeted by winds of intense ferocity. You can be sure of this: it will not be nice.

But clip we must. Especially as Little John (clipper in chief) is off on his holidays next week.

I catch the ponies and shout the tups down to the pens and that much, at least, goes to plan.

But John seems to have forgotten that Manxes do not like being rounded up by a dog and Freddie's little dog Floss is no match for them. Usually – and within the twenty-four-hour notice period that is obligatory for keepers of primitive sheep who are all patient souls – I can get them in with a few lusty hollers and a bucket. But instead, today, we chase them several unsatisfactory circuits of both fields, the dog gets bruised, John loses his temper and a lot of swearing goes on. In the end, we capture them: John resorts to rugby tackling and carrying two of them and the rest are herded, reluctantly. They glare at us.

With modern clippers that run off a tractor battery, John shears two and then pauses to assess the situation.

'How many of these are there, father?'

Freddie looks – well, sheepish, really.

'You said there were eleven. There's more . . .'

'Fifteen,' I say brightly, rolling another fleece the wrong way out. Which happens to be the right way out for rolling fleeces but that's a technicality that, if you live in suburbia, you scarcely need bother about.

One wriggles, with near-death consequences. 'Keep bleeping still. I am bleeping near your bleeping throat and I have bleeping sharp blades . . .'

Three from the end, another wriggles and there is an unscheduled case of castration – well, almost. 'Mind his willy,' shrieks Freddie. I mutter something about procedures like this having enormous potential vis-à-vis NHS savings on vasectomies and John carries on, deft and swift and bent. The rain threatens, big blobby drops falling, then stopping just as abruptly. The sky is leaden and the bald sheep are very, very cold.

After another half-hour, two brief rests, a phone call from his wife, a discussion about whether takeaways are justifiable when both partners work (we decide yes, on balance, especially for curries) and a comprehensive denunciation of the new blue bag refuse system ('I said, "If you leave me two bleeping blue bags every bleeping week then I'll use them but if you don't I'm sticking my rubbish in black bleeping bags and I expect you to bleeping empty them," that's what I said . . .') we are done. A heap of soft brown fleece is laid out on a blue plastic sheet in the shed – unrelated to the blue placcy bags, obviously – John dashes off for his dinner and Blossom, now back in the field, is utterly distraught. She cannot comprehend where the sheep have gone and why there is a herd of black pygmy goats in their stead.

It is very perplexing being a pony, at times.

*   *   *

Tess has been licking her stitches and set up an infection: nothing too desperate, but messy. We decide to take her back to the vet sooner than arranged and, indeed, a course of antibiotic and anti-inflammatory pills are prescribed.

Now Gyp is the kind of dog who is easily conned and so greedy that she scoffs whatever is offered and worries about what it might have been afterwards. Tess is different: she eyes everything, even Pedigree Chum Pouch Packs, as if you may be actively seeking to poison her (you are not; think of the vets' bills such reckless-ness would incur) and eats daintily: not a dog for duping.

Tess aspires to feline levels of suspicion. Just in case you ever need it, here is how you give a cat a pill.

- Pick up your little cat and cradle it lovingly in the crook of your left arm as you would a baby. With the other hand, gently squeeze the cat's cheeks, and when the cat opens its mouth pop pill into its waiting jaws.

- Wait until the cat spits the pill out, pick it up and start again.

- Force the cat's jaws open, pushing the pill to back of mouth with right forefinger. Hold the feline's sharp teeth together for a count of ten, if you can bear the pain that long. The cat will spit this out too so summon reinforcements.

- Get spouse to wrap cat in large towel and lie on cat. Put pill in end of drinking straw, force cat's mouth open and blow down drinking straw.

- Retrieve cat from neighbour's shed. Try again. Fail. Open a beer. Drink beer. Fetch bottle of Scotch. Pour shot, drink. Apply cold compress to cheek and check records for date of last tetanus jab. Apply whisky compress to cheek to disinfect. Toss back another shot.

- Tie the little bastard's front paws to rear paws with baler twine and secure to chair leg, and using falconry gauntlets shove pill into cat's mouth followed by two pints of water to wash it down.

- Consume remainder of Scotch. Take mutant cat to RSPCA and ring local pet shop to see if they have any hamsters.

How to give a dog a pill: wrap it in cheese.

You can't get away with wrapping Tess's pills in any old cheese, though: it has to be Garstang Blue. And I am not joking. I tried her with Feta, Cotherstone, smoked Applewood (came close), but in the end it was only the Garstang that had a 100 per cent success rate.

The infection had another side effect: Cover Dog had to endure the monumental indignity of walking about with a lampshade on her head. Unlike the tups, who are aware of the width of their antlers, Tess fails to get the hang of this and keeps crashing into furniture; if we had any priceless ornaments on low tables they would have been reduced to smithereens – whatever they are – by now. We haven't, so they're not. But bear it in mind, for future reference, won't you?

I forbore to decorate the clear plastic hood with my glass-painting kit, in spite of being fairly confident I could knock up something vaguely Pre-Raphaelite in design and colour scheme, as it seemed cruel to mock the afflicted dog. Tess knew she looked a prat in the ridiculous headgear so I consoled her with the tale of Puffin, the unspeakably hideous hound we kept as a pet when I was in Mixed Infants. Puffin had to wear a real lampshade. Granny found just the thing at the school rummage sale – pink, satin, pleated; tasteful it wasn't, nor elegant, nor appropriate, just very, very silly.

This morning, I fed Lazarus, put him in the runabout pen and had set off up the field armed with bucket, feed, stick and carrots for Bob when a small nose jabbed furiously at the back of my knee. 'Blaaaa . . .' I turned and there was Lazarus. This was strange because I had sheep-proofed (insofar as anything can ever be sheep-proof with Manxes) the pen. My proofing had kept the little beast contained for the past few days. The gap eluded me. Everything seemed sound. No rolled troughs leaving telltale little gaps under the pig netting fence. No holes in the wire. No dodgy knotting on the baler twine gate fastening. No gaps, nothing. I set off again, armed with bucket, feed . . . we've done all this once, haven't we? I snuk a glance over my left shoulder.

Nothing. No lamb trotting in my wake. Must have been a mirage, then. I almost reached the water trough and then stab, jab, poke in the back of the knee and 'Blaaaaa . . .'

I picked him up, turned round and finally saw how he had escaped. He had probably not noticed it before, the slender space between the stone wall and next door's fence. The place I hadn't checked because I had assumed it wasn't there.

Bright little sheep, Lazarus.

A spare straining post filled the gap nicely, but just to make assurance doubly sure I tied a hurdle to the fence and wedged it with a stone too. Can't be too careful if he's getting this smart.

Mind you, certain as I was that the perimeter fencing was impenetrable, the first thing I did on returning from Penrith was to check if I had still got a lamb in the pen.

It's time to creosote the gates again. I do this every summer just before we go on holiday. It means I depart guilt-free and exhausted. This year, though, I am not supposed to use creosote. Why? Because DEFRA, in their supreme wisdom, have banned it. Come to think of it, it's been a while since I had a pop at the EU, DEFRA and their ilk. What is an ilk, by the way? Do I have one? Do you? Are there different sizes? Anyway, I would not like the authorities to think I had taken my eye off the ball, so here goes.

First on the list of regulatory bodies that have got on my wick lately is the DTI. Perhaps I should be grateful that we bother to have a Department of Trade and

Industry at all when precious little is left of either in our modern service economy, but I am none too thrilled that the mandarins of the DTI have suddenly and inexplicably decided to deprive me of the occasional creosote fix. Why have they suddenly taken against creosote? I have daubed the stuff on my gates, fencing, hurdles and face for two decades and it hasn't done me any lasting damage. Unless, of course, you know differently . . .

At the very least, they owe it to me to tell me what it is that hasn't affected me at all.

The DTI directive propped up on my desk 'twixt promotional wormer coffee mug and snow globe of the White House – an odd juxtaposition but an effective one, directive support-wise – enquires: 'Do you have any Creosote at home?' I'm not answering that on the grounds that it could incriminate me. I might have been secretly stockpiling it. And then again, I might not.

The DTI bang on a bit about water-extractable phenols and benzo-alpha-pyrene in less than 0.0005 per cent concentration by mass but they say nothing about how appealingly cheap creosote is, nor do they mention the satisfyingly aromatic bouquet that infuses a newly dressed gate. I think they are missing out, myself.

But that's it: curtains for creosote. Perhaps we can look forward to the agricultural landscape being magically enhanced as a result. All that drab brown will go and be replaced by fancy rainbow hues of Winter Holly green gates, Blue Tit blue (obviously) doors and Sellafield Neon (only joking) fencing, though

I expect the wood treatment manufacturers to continue pandering to beige people with a shade of Pulverised Hedgehog.

Me, I might just sneak out at night with my stockpiled creosote. If there is such stuff, of course. I'm not saying.

I suspect that the DTI was also the agency behind the excision of Zebo from the national shopping list. Zebo is – was – the stuff I used to clean the grate. It was By Appointment to HM the Queen, too. Whatever does she use now to give the regal fires a decent blacking?

Next, predictably, step forward DEFRA. You lot, I see, have been amusing yourselves with new regulations regarding on-farm burial, insisting that all dead livestock (what a peculiar phrase that is) are ferried away by registered operatives and incinerated. There used to be a saying round these parts: when you buy a sheep, buy a shovel at the same time. DEFRA wants us to toss that nugget of received wisdom down the toilet bowl of history; in a headlong dash to appease the Europrats it has been decreed that we shall no longer be permitted to bury our dead in peace. Sheep, pigs or cattle, that is. Interestingly, there are no rules about pet canaries, dogs, cats or gerbils (they fit neatly into a redundant Tesco bag and can be put out with the bins, I guess), and as for people, well, just hang on, I'm coming to them in a minute.

But isn't burying animal carcasses exactly what DEFRA insisted upon during the foot and mouth crisis, or did I get that bit wrong . . . ?

And now to humans: how do DEFRA's latest raft

of spiffing wheezes sit with the increasingly popular option of burying people in woodland burial sites? Weirdly, it seems that you can bury humans on farms, but not farm animals. DEFRA might, of course, be planning to scour the countryside for roadkill crows and cart them away for incineration too, thereby depriving scavengers like crows of food. That sounds very like cannibalism to me, but the legal eagles that keep a beady eye on their neighbours wheeling about among the clouds tell me that crows are just very environmentally aware birds keen on recycling.

My objection to the double-tagging regulations is rather more serious. I hate *doing* tagging, you see; singular or plural, it is a deeply distasteful practice. I would much rather drink neat creosote than fire that tagging gun, listen to its softly sickening crunch and watch the inevitable droplet of guilt-inducing blood leak from a young ear. And now I must do exactly that, twice.

At first glance, I thought that another of DEFRA's decrees requiring horses – or more precisely 'domestic animals of equine or asinine species, or crossbreds of those species' – to hold passports seemed quite a good idea. It might tighten things up on live shipments, I thought. It might make a difference in tracing lost and stolen animals. Neither is likely in practice. All passports will do is add to bureaucracy and raise funds for the government: that's tax to you and me.

Yet quite why DEFRA are insisting on passports for all equines remains unclear, because elsewhere in Europe EU Directive 90/427 requires a passport only for 'registered' equidae, or, to put it another way, those entered or eligible for entry in a studbook. So, if Bob and Blossom lived in Spain instead of in my back field, only Bob would need a passport as he is a registered Welsh Section C pony. Blossom, being the bastard product of an illicit liaison, would not. As it is, I have coughed up a tenner for hers already. Blossom's pass-port is blue, not leather clad like those lovely old British ones used to be but plastic; at least the drawing of her doesn't make her look like something on a police WANTED poster like the photo on my own passport.

Just when I thought I had reached and breached the final frontiers of stupidity, I came across Section 4.8 of EN71 Part 8 which states that 'When measured from the ground to any sitting or standing position, the maximum free height of all carousels and rocking activity toys shall not exceed 600 mm'. And, barmily, that particular bit of Brussels nannying outlaws some rocking horses.

What a very strange world we live in.

And while we're on the subject of 'stupid', does it worry you at all that you have spent several minutes reading the rantings of a geek who collects such point-less facts together and writes about them? I really think it should.

It's Saturday and Mickey is competing at Cumwhinton Horse Trials. Usually we can't get to Saturday trials as I am too busy cleaning and baking for incoming cottage guests but today it is so close to home that we can just manage it – if we scurry. I am thankful that Sarah has managed to get an early running time because enough rain came down last night to power a couple of hydro-electric plants. Lake Geneva has migrated to the road by Haningbrow Wood on the way down to Armathwaite and the trees are glistening with still-wetness. At least being one of the first to go across country means that she should get round before it cuts up. If she gets round . . . it is now that I realise that I am rubbish at this owner lark. All I want is my precious boy – who looks the real deal, shining, muscular and neatly plaited – to come back safe, sound and in one piece. His dressage score looks OK – 42 penalties. He has had a brick out of the last wall in the show jumping and is standing, taking it all in, at the back of the wagon, sweat rug slung insouciantly across his back. One cool dude. All he needs is a set of wraparound shades and a pair of Italian loafers. No, make that two pairs of Italian loafers.

I have my new digital camera and I manage a more or less still of them in the practice ring, though I amputate Mickey's legs. He would be no use like this across country. I watch them practise a bit – he still appears to be half asleep – and then they are off. Over the first. My photographic gene has shut down altogether: I don't hear any sort of click. Never mind – listen to the commentary . . . Mikolai and Sarah Linton are over the third . . . on – and off – the bank . . . through the water . . . over the beehives . . . nineteen fences later I am standing watching them jump the flower beds, the last, and missing the photo-opportunity again. A new camera and a moving target was just too much of a challenge first time out. It hardly matters. I weep with excitement, relief, pride. 'You clever, clever boy . . .' I tell him. He's dripping, now, part water part sweat, and Sarah is walking by him with a smile that will take days to fade, a smile as wide as a San Franciscan freeway.

We all wash him down, pander to his whims and feed him with an unhealthy quantity of Polos. Back to reality, and the cottage cleaning for me and a bit of light duty spectating for Sarah and her long-suffering mum.

That's our little team: one horse, one rider, a band of willing hands and well-wishers and a slot in someone else's big wagon for the day. Things are rather different next door. It's a horsebox that will have cost – conservative estimate, this – over £100,000. It has a solarium for the horses, central heating for the humans and every conceivable form of off-course entertainment system. And a satellite for something, though I'm not sure what. It is built to transport about six full-size horses in considerable luxury. And here's the thing: they are competing in the same class as us.

Some, as Orwell rightly noted, are more equal than others.

Sunday and we are jump judging today, leaving the new guests in the cottage watching the garden birds feeding on the bag of goodies that last week's visitors thoughtfully left behind. Malcolm's home-made-in-an-afternoon-from-rubbish bird table has been appreciatively received.

As with all horse trials, it is an early start for the

briefing. The usual suspects are all there: local worthies having a day off, retired colonels coming out to play and retired competitors putting a bit back into their sport, along with the young generation of volunteers being initiated into the mysteries of horse trials policing. It is run under the same rules as Brougham was in April, with the same myriad regulations about everything – what to do, what not to do and what order to do it all in. I hope fervently that we shan't need to use the pink accident report forms, especially not the box – and this really is no joke – with FATAL written in it. Horse trials have suffered several fatalities in recent years, and in spite of massive efforts to make the sport less dangerous, riding a horse at speed over solid obstacles remains a huge adrenalin rush and can never be entirely safe. Nor should it be since that very risk is part of its allure.

We drive off to Fence 6, the corner question, to which I am absolutely sure some will fail to find an answer. The first thing any woman needs to do is stake out the nearest loo; it's within comfortable pee-break distance, just a few yards. I love the mobile unisex loos at events. They are always a wildly luminous colour and they have brilliant notices inside. I especially like the one for the chaps: 'Urinal – Aim Here' with usually a couple of green aggressively pine-scented tablets to act as targets.

Fence 6 is huge: triangular, wide, imposing and terrifying. I still think and walk as if I would have to ride it and the jumps still strike terror into my soul. Malcolm thinks one or two will take the right-hand, more direct route. I disagree. 'Why? It's so much wider and more difficult – no-one will be daft enough to make it more complicated than they need to.'

The first competitor takes the right-hand line. And that explains much, I fear, about my own competitive career.

The Novice is a big track and rides well. When I was competing in my teens, Novice was much, much smaller than this; in those days, a fence this size would have been Open Intermediate standard. So much have things altered that the eventing establishment has needed to bring in first Pre-Novice and now Intro so as not to frighten off the horses and the riders. More than four hundred competitors are here this weekend and, happily, our fence rides brilliantly. Mostly, they negotiate it with a whoop or a growl and an occasional 'Oh shit.' One even manages 'Oh shit – *good* boy – sorry for the swearing – byee' as she heads off over the hill. Magic.

The special prize for Meaningful Social Intercourse While Very Mobile Indeed went to celebrated international competitor Ian Stark who, recognising an acquaintance, called out, 'Hello, how are you?' though I noticed he didn't hang around for a reply. And I thought only women could multi-task.

Occasional hiccups occur. There's the obligatory stroppy mare: first the horse stops, then the stroppy mare (you didn't think I'd use that sort of terminology about a *horse*, did you?) shows her horse the fence pretty overtly contrary to the rules, circles away and stops again. She grimaces and asks 'How do I get out?' before attempting it from the other side and failing again. By this time the course controller is in a lather. 'Omigod,' he says to our neighbours over their radio, 'she won't like it. She won't like it at all.' It appears that this lady's reputation precedes her.

Another jumps the wrong fence. Beautifully. She has no idea whatever that she has taken the wrong course and comes back later to take issue with us, but very politely. Her husband suggests that we have been mistaken. She is too well brought up to be anything other than politely adamant, but she still insists that she jumped the correct fence. 'If you have any doubt at all,' I explain, 'go and see Douglas because he and the official photographer just happened to be here when you jumped the fence. So there's photographic evidence.' It is her turn to say Omigod.

We part the best of friends, she berating herself for being 'utterly pathetic'.

Later, we have that awful situation of a potential pile-up. Not because of any accident or fall, just one travelling so slowly that the one behind is close enough to overtake – and we are only at Fence 6, remember. The one in front is totally oblivious. The catching-up competitor yells. The one in front appears to wake up and adjust the steering but looks slightly affronted as she is overtaken. There is just not time for a speech along the lines of 'I am approaching from the rear at a speed of approximately 26.8 miles per hour. I am heftily booted and will leave a permanent imprint in the event of impact, which will have a percussive force of . . .' You just yell.

Boys, girls, grown men and women, old-enough-to-know-better equestrians, Upper Class Twits of the Year, wimps, blondes, brunettes and redheads, bespectacled ingrate (you know who you are) and the terminally lost, all have my undiluted admiration: brave Corinthians every one.

Off stage, Gyp and Tess are receiving their public, wagging tails and licking all comers appreciatively. We also share our very good lunch with them: little bits of fine pork pie, organic unpasteurised Lancashire cheese roll, some shortbread stars and a bit of raspberry cake (for the fibre content). They don't think much of the fancy Willow water with its natural salicin, though, even if Clarissa Dickson Wright does reckon it cured her zits.

The clouds have passed and the sun is shining. Blue sky, green grass, rippling horseflesh, Olympian competitors and all for two measly quid. Who says going out is expensive?

At the end of the day, I was aware that I had told anyone who showed the slightest interest that Mickey had jumped clear yesterday and that we would make an event horse of him yet. I expect they all felt like giving me a good kicking.

So, we have a new Countryside Code. I like Codes. There was the Enigma Code (never did work that one out), the Green Cross Code (created to prevent infants from getting hedgehogged) and now this brevity-is-the-soul-of-wit Countryside one.

I could recite the original version – a wonderful gobbet of liturgy – with an easy eloquence when I was in the Brownies. I was a rotten Brownie; our Brown Owl aka Brown Cow, for reasons never satisfactorily explained, made us travel fifteen metres (yards, in those days) using upturned Lyle's golden syrup tins as stepping stones. I was hopeless at it.

The old Countryside Code kicked off with: *Enjoy the countryside and respect its life and work*. Now I don't wish to be pedantic over semantics, nor do I wish to be a poet, but that's nonsense. The countryside doesn't work, per se; the people in it do. But I'll gloss over that (Lynne) Trussist annoyance and go to the next point: *Guard against all risk of fire*. No bang-bang-pop-pop-you're-dead-I'm-not fireworks, then, but probably no real need to go out in flame-resistant suits either. That would be too silly – almost as silly as travelling fifteen metres on upturned golden syrup tins.

Fasten all gates came next. Yes, and I wish those two Carlisle pensioners I came across in the bottom field several years ago had heeded this one. They were busily stashing a startling collection of mushroom-stuffed plastic bags into their family saloon. The gentleman resembled a bandicoot that had breakfasted on a brace of lemons instead of some tasty invertebrates and assured me that the fungoid loot in the boot was all for their own personal use. Of course it was: silly me. He added peevishly that no-one had bothered him when he had been mushrooming before. I bothered, though; less because of the mushrooms, his addictions and afflictions being his own affair, than because I arrived just in time to witness a neat crocodile of my Mule ewes passing through the gate he had left so invitingly open.

The ewes were advancing upon a Johnny-foreigner tup. And they had plans, I could tell. The pensioners didn't even offer to help me sort out the incipient orgy but flung down the boot lid and made off in a squeal of brakes and a cloud of dust.

I had some funny-looking lambs the next spring.

There are times, though, when gates are left open for a good reason; the route to a water trough being an obvious example. So I prefer the new Countryside Code non-intervention policy on this one. It says: *Leave gates and property as you find them.* If the property concerned is a piece of deadstock in a machinery graveyard and has been stranded in the middle of a field for more than two decades, it deserves to rest in peace, providing sanctuary for a colony of voles, woodlice and other things with an excessive number of hairy little legs.

The instruction to *keep dogs under close control* remains exactly the same in the new Code. I don't worry too much on this score since G&T don't even mildly irritate sheep, much less worry them – not that they have been out much lately as they've both been too poorly.

The Right to Roam has not altered the necessity to *keep to public paths across farmland*, but this is not specified in the new Code. It would be nice to think that the cataclysmic 'chaos and conflict' that some land-owners are predicting will not come to pass, but there could be some confusion about where farmland ends and upland or moorland starts. And people will still get lost.

The old Code urged walkers to *use gates and stiles to*

cross fences, hedges and walls, but now that we are left to our own devices, don't forget how handy crampons can be for those awkward uphill moments. And for the really extreme scenarios, there are always hang-gliders and microlights, but don't hold me responsible if it all goes bandy and you end up dangling from an ash tree making conversation with a crow.

To my mind, the most important rule of either code is: *Take your litter home*. Here's why. One of our fields leads off a lay-by. All sorts of things go on in that lay-by apart from accessing fields, and although it would be a shame to rat on most of those irresponsible I'll make one exception. He had had his lunch – chips. Perhaps fish, perhaps jumbo battered sausage too, I don't know. He settled down in lumbar-supported comfort for a post-prandial listen to Radio 1, screwed up the paper wrapping and tossed it out of the window. Rather unfortunately for Chip Man, I was shepherding just out of his line of vision; I crept alongside, picked up the crumpled orb and drawing on my time in the Ladies' Cricket team (I was rather better at cricket than balancing on golden syrup tins) fired it right back in, catching him neatly on the jaw and, to my considerable satisfaction, frightening the wits out of him.

I expect he took his litter home next time he finished his lunch.

The bits about keeping water clean, protecting plants and animals, taking special care on country roads and making no unnecessary noise have been excised from the new Countryside Code altogether, in a drive for brevity rather than clarity. All of it, of course, is based on the premise that most of the public are reasonably sensible. This could prove dangerous, perhaps, but not as reckless as the latest defence cuts which effectively reduce the British navy to the same strength as the Swiss one.

Whoever thought that one up is guilty of gross stupidity. That's one hundred and forty-four times as stupid as normal, to be absolutely specific.

The Back Quack came out to Blossom tonight. Blossom had a funny five minutes in the trap last week and seemed to be rather cautious about going downhill when I rode her so it was reasonable to assume that she had put something out in her back. A vet will tell you to rest the horse and for some things that's just fine, but it won't cure or unlock a trapped nerve. So Bob the Back Man came to call.

Now Bob is one of those characters who, if he didn't exist, could not possibly be invented. He was born at Dufton on the east fellside and started his working life as a joiner making window frames. Bob didn't take to window frame making.

'My boss Mr Rudd looked at one frame I'd made and asked me if I had ever seen a worse window frame in my life. I said, "Oh, yes, there's another three on the bench *far* worse than that . . ."'

So he went to work at British Rail instead, where one of his jobs was to wave a flag to stop a train in the case of emergency. He wasn't much good at this either. 'I failed my test as a relayer the first time. The examiner asked me what steps I would take to stop a train travelling towards me at a hundred miles an hour. I said f— great big ones . . .'

After that he turned to country and western singing and is still in demand, though the 'horse job', as he calls it, takes up too much time to allow him to do much singing now. His lifelong fascination with horses developed into something more like an all-consuming passion when he met and married Margaret, herself an accomplished rider; they bought a 'funny little horse called Mockaite who had a bad back. We paid six hundred pounds for him at the sales and everyone said that the only way he'd travel two miles would be in the back of a furniture van.' By now Bob had realised that he had healing hands and Mockaite, one of his early patients, won at Carlisle.

Word spreads fast in the horse world, and now that the benefits of alternative therapies are widely acknowledged Bob is busier than ever, using his combination of healing fingertips and magnetic field treatments. Blossom, needless to say, had put a couple of small things out of alignment in her back. Bob tweaked and thumped and had a little chat with her to apologise and tell her that it was all for the best and then she was fine. Then he did my shoulder. And then he did Malcolm's neck. A sort of buy one, get two free deal.

Blossom gets a rest of sorts now anyway. We, you see, are off on holiday.

August

A favourite month for visitors. Sometimes, if I have a suitable pony, I'll stick a visiting child on it and walk them round the field. One young bronco-buster announced that he wanted to ride like a cowboy, and before I could stop him he booted the pony in the ribs. The pony took matters into his own hooves and the circuit of the field was accomplished with childish knuckles shining white and the small rider fully cognisant of just what ponies, impolitely dealt with, can do. Another little boy, this one totally in touch with his limitations, just squealed, 'It's freaky up 'ere. Get me off!' Imagine what might happen with unsuitable ponies . . .

OH DEAR. HOMO unintelligentsius holidayus. He's everywhere and you can tell the Brits a mile off. They're the ones with the sock and sandal ensemble; beige, usually. Still, I suppose we should be grateful for small mercies: the days of the knotted handkerchief are over.

Malcolm, meanwhile, is panicking. The plane is not on the Departures board yet. We shall, he says, be delayed; it is just a question of by how many hours. Or days. I trot obediently to the desk to explore the whys and wherefores, he being inexplicably tongue-tied on such matters. I am assured that it landed on time but Malcolm doesn't believe me. We go in search of liquid refreshment. Just as I get to the front of the queue, the guy behind the counter goes for his lunch (it's eleven twenty for heaven's sake) and I am left to alternate with the neighbouring queue. It is times like this that I wish I had done some assertiveness training.

I gather up an unsatisfactory cappuccino: the chocolate shakings have all stuck to the lid but it is hot and creamy. Malcolm sips his juice and I reflect that things could be worse: I could be going on holiday with the family sitting opposite us. They remind me of the Osbournes. Father is an ageing rocker with long curls; mum is shrink-wrapped in jeans; number one son has earplugs in and is in nodding-dog mode; number two son, similarly wired, is making a strange noise that sounds like a muffled pneumatic drill and number three son stares absently into space. A dynamic lot. 'He's bloody idle that son of yours. Miserable little git, why didn't he bring me a bag-wet?' I'm sure they'll have a lovely time. I'm just very glad that I'm not going wherever they are. I'm not, surely . . .

Our flight is called and for all that I have been apprehensive about flying Croatia Airlines we board a brand new Airbus and settle into seats with agreeable legroom. This is not going to be so bad after all. As long as the weather forecasts I found on t'internet – four, possibly five days of unremitting thunderstorms – are wrong, it could be fun. If they are right, I will wish I had stayed at home.

Dubrovnik, two and a bit hours later and just about on time, is hot. Very, very hot. We step on to an airstrip at the base of a mountain range – I am certain that the pilot let the wheels down a little too quickly. I'm sure he should have hung on a nano-second or two. But here we are, safely delivered. An enormous bus for so few passengers takes us along the newly constructed highway; there is little evidence of the bloody civil war that tore the heart and soul out of this land so recently. Instead, it is a peaceful landscape of timeless beauty: mountains, pine woods to our right and a shimmering sea to our left. We disembark at the pedestrian gate and follow our portered luggage up fifty-two steps, immediately under the mighty, mediaeval city wall. The apartment is cosy, clean and neat; there's a welcome pack of everything we could possibly need and an enchanting terrace with a view over Dubrovnik's extra-ordinary roofscape to the Adriatic glistening beyond.

I think we have done rather well.

As long as the thunderstorms keep away, of course.

The local beer is good stuff, the ice cream even better; local dried ham, fat peaches so ripe that you have to eat them with your chin stuck out and your knees apart to avoid dripping the juice down your clothes, dried sweet figs and colossal tomatoes mean that cook-ing is completely unnecessary. Pizza is everywhere, the legacy of Italy's proximity and influence; fish is plentiful if not exactly cheap and the sunshine continuous. We sit, that first evening, on the terrace, marvelling at the scene and enjoying the unexpected quiet. We had thought that being in the old city carried a risk of noise but the walls afford protection from the lunatic excesses of motorists beyond and our eyrie is silent save for the birds swooping overhead and church bells chiming the hour. Even the noise from the Stradun filters away in an unseen vortex on the night air.

The next few days pass in a pleasant daze of sunshine, total immersion in books or water, swimming from rocks in translucent sea and sightseeing. We walk the wall, mesmerised by buildings with no apparent means of support and even less foundation; by tiny windows set at crazy angles and by the aerial mosaic of old and new terracotta tiles. There is one dramatic and terrifying electric storm but it is at night. God's fireworks illuminate the sky, thunderclaps rattle the apartment, but by morning the air is cleared and the heat is no longer dense and humid but clear and dry. The Adriatic glimmers an impossible blue; shafts of sunlight ricocheting off the waves dance in endless, perfect rhythm. Yachts, playthings of pop stars and moguls, bob and buck on the distant horizon and a cruise ship inches into port each day. Gigantic floating tower blocks of bombastic, balconied brashness alternate with sleek white hulls, though whether these smaller vessels are ships of understated elitism or just plain stuffed to the gunwales, all inside cabins, titchy portholes and bunk beds, is hard to define. We don't care much, one way or the other; we just watch them, towering above the fishing boats and ferries and jet skis in the harbour.

We travel into the interior to Mostar and the newly reconstructed bridge. Men jump off this bridge, not as I thought to mark some particular event, some celebration or festival, but all the time. I think they are mad. It's quite a drop. Oddly, for a country that specialises in dried ham, we see no pigs, no sheep – they're up in the mountains and since the mountains constitute about 40 per cent of Croatia's land mass there are probably quite a lot of them – and only an occasional cow. It's always possible that Croatia breeds a special sort of invisible pig, because ham is plentiful. We pause on the way back to Dubrovnik for a pee-stop and fall into conversation with a couple who have been watching the local traffic cops booking anyone who exceeds the speed limit; Malcolm says that this level of assiduity is proof positive that our interlocutor is a retired traffic policeman. I am unconvinced: such circumstantial evidence would never stand up in court.

If you've never been, here is everything you need to know if you are visiting Dubrovnik for the first time.

• Sunshine bleaches your hair. Even the grey bits.

• Grasshoppers look mean but they don't eat much. They have an indeterminate number of particularly bendy and angular legs; they use these to bound, in a single leap, from the terrace to behind the kitchen door where they lie low for a while. You may only find out that there is a grasshopper there as you step back to shut the door, in your bare feet . . . an unpleasant crunching sound under your left heel will alert you to its presence. You are permitted to expel a small squeal and a single swear word at this point. Grasshoppers can hang upside down on parasols and threaten to drop on your head but they prefer hanging to descending sans parachute so, generally, they stay where they are. Anyway, even if they don't and drop into your lunch they will not eat much, as noted above, and what they do consume will not be anything you would kill for.

- Squid does not always taste like elastic bands. If you go to a decent fish restaurant it will, I promise, taste surprisingly good.

- Pizza is obtainable across Dubrovnik at any time of the day or night and is universally sound. Figs are excellent but too many, however good they taste, are not a clever idea.

- No matter which way you walk the wall, every way is up. Take Radox if your accommodation has a bath, horse liniment if it does not.

We had a wonderful week. What's more, we returned chipper and revitalised.

The next day we go and pay the ransom to release G&T from custody. They are quite pleased to see us. Not delighted or thrilled or anything like that, but quite pleased. It is a funny thing, this whole business of putting dogs in kennels. When I take them, the little beasts trot off without so much as a backward glance. They are far too busy ingratiating themselves with the girls who will look after them for the forthcoming week to bother with me. Once I returned home in tears. 'They don't love me,' I wailed.

'Don't be stupid,' said Malcolm. 'You'd be more upset if they howled and went off their food and pined when they were left.'

Yes, but I would like to feel that they might miss me just a teeny weeny bit.

Anyway, now they are home and, as usual, they smell lovely. They get sprayed with something called Johnson's Velvet Coat when they leave: Kenzo Flowers for dogs.

'Just make sure you put that PRESS badge on carefully – so that it is identification and not an invitation,' said my husband in tones suggesting that, like a lurcher, I could not be trusted off a lead at Lowther. I felt very legit wearing my painstakingly positioned PRESS badge and trotted off to the *Cumbria Life* stand to sign hardback copies of *The Funny Farm*, passing several extremely handsome dogs and a worryingly disproportionate number of extremely ugly humans en route.

I have two signatures, these days: an informal one for the books and a longer one – Jacqueline E Moffat, suggesting an individual with both tangible assets and gravitas – for the cheque book. I've already used an entire Parker ink cartridge for book signing but none for the cheques lately.

I found I could sign the books OK but I ran into

awful difficulties with the selling process, having no idea how to navigate the transactions and even less clue when something called 'payment processing' was involved. I tried to do a deal with one purchaser who had neither change nor cheque card. 'That's OK,' I said, 'I'll take the dog' – a particularly attractive spaniel puppy – 'instead.' Negotiations faltered when they said no, really, they'd rather hang on to the dog but how did I feel about their child. I told them I don't do children and they found their cheque book.

Lowther was a very jolly time. It is enormous fun making new friends and meeting up with old ones and calling it 'work': it's job satisfaction with bells, whistles and party poppers on.

While we have been away on holiday new guests Jenny and Rob have arrived for a holiday in the cottage. Police officers taking time out from the assembled villains of their home county, they are relaxing with their dogs, a Rottweiler and a Dobermann. If you think it would be difficult to relax with a Rottie and a Dobie, then you really ought to meet Alice and Billy.

Billy is the world's only ginger Dobermann. Probably not quite pedigree, then. This did nothing to deter his doting owner from paying full pedigree price, though he adds in mitigation, 'I was drunk when I bought him.' Rob's sister said that the dog was part Dobermann part guinea pig and within the space of a very few weeks Billy the dog, Dobermann or not, had racked up an account at the vet's so large that Rob was considering selling off body parts to meet the cost. Jenny, however, hoped there might be another way round this dilemma. Rob was unwilling to concede defeat so early on and his fiancée was easily led, so, in an effort to subsume Billy's more manic tendencies, they went out and bought him a companion. Enter Alice, a Rottie with a white bib, also known as the Runtweiler. 'We will buy a proper dog one day,' they aver, a tad unconvincingly. Alice, the inspiration behind (and I use that preposition advisedly) the song 'Shakin' that arse, shakin' that arse' has an altogether amazing arse; she is the only dog – never mind dog, she's the only creature – I have ever seen who manages to waddle athletically. Usually that's a contradiction in terms, but not in Alice's case.

I was just relieved to find that they had settled in so well in our absence. I rarely do this, go AWOL when visitors are due, because these days lobbing the keys in the general direction of a holidaymaker is no longer enough. Discerning guests expect high standards of hospitality and cleanliness and quite justifiably so. So if you go away yourself, make sure that you leave it all in a safe pair of hands, ideally belonging to someone who shares your own irrational phobia of other people's pubic hair.

All was well. Needn't have fretted after all.

Major panic today, when a visitor to Rowfoot leaves his car keys in the ignition and his dog somehow – God knows how – locks the doors. Spare keys? No, they are in Jersey, with his son. Silly me. I should have realised that. There is much to-ing and fro-ing, ringing and waiting and looking for wire coathangers. Rob and

Jenny's police – and therefore criminal – connections inspire them to toss into the conversation that they might be able to break in, given a minute or two, a bit of string and that coathanger. There's a snag here: the English Tourist Council have something against wire coathangers and insist that none lurk in their approved establishments. They just don't understand how vital they can be if you need to break into a car at short notice, I think. It turns out that Vauxhall have taken steps to prevent ingressions by coathanger, string, screwdriver, monkey wrench or chisel anyway (we know, we tried them all), and eventually the small side window has to be smashed. It would have been cheaper and quicker to call out the AA but at least the dog is none the worse for the ordeal. Given any longer she might have been; overheated dogs can con you into thinking that there is nothing wrong, they are just having a doze. Unlike people with pubic hair phobias, they don't get distressed and start fretting, they just become comatose and sleep their way out of the world.

You really don't want that to happen, so leave them a window open. Better still, don't give your spare car keys to anyone who is going to Jersey on holiday.

It is the monsoon season in Ainstable. We have a lost tributary of the River Eden running down the road outside the house. I am glad we live on a hill.

It is still raining.

More rain.

*   *   *

Some places in Cumbria have had four inches of rain in the last twenty-four hours (no, I don't know what that is in centimetres, and I'm not going to look it up). That is a lot of water. The dip between us and Faugh Head is under nearly a foot of water. At least one car has not made it up the hill.

It is a beautiful morning. No rain. This is good because today we are going to Ravenglass to have lunch with Patrick and that means a foray to the west of the county, bypassing places with weird spelling and even odder pronunciation like Aspatria pronounced Spay-tree-uh and Torpenhow which, you will recall, is pronounced Trappenna. Quirky local pronunciation may account for why a great many people get lost in Cumbria, but it can't explain the case of the guest invited to a wedding at Armathwaite – nice do, posh marquee on the lawn and all. He failed to consult his map and booked himself and his good lady into Armathwaite Hall Hotel in anticipation of a fine night's kip not too far from the shenanigans. So far so good. Only when he arrived at Armathwaite did he discover that Armathwaite Hall was located at Bassenthwaite some thirty-odd miles away. The hotel is ritzy and smart, but it is said that his taxi fare back to Bassenthwaite in the early hours of a morning significantly exceeded his accommodation bill.

As if to amplify the joke, a major horse trial (three found guilty, one remanded in custody and another still in the lake) was held at Armathwaite Hall. The equestrian media unhelpfully referred to it as 'Armathwaite Horse

Trials' and this may have contributed to the number of horseboxes the size of Australian road trains that had awful difficulty turning round on the road between the pub and the post office in Armathwaite. A few serious stand-offs occurred when one road train met another coming the other way, but no bloodshed ensued. But be warned: all may not be as it appears, wherever you might think you are going.

Although we know where we are going and it's not to Aspatria or Torpenhow, there's a tale about Aspatria that needs telling. Once upon a time, in the days when the trains that stopped there had three classes, the stationmaster would approach the first-class carriages and announce, in suitably clipped tones: 'Uh-spay-tree-uh, first-class passengers please alight here.' Progressing to second class he would shout 'Spay-tree-uh', and when he reached the plebs in third class he just hollered, 'Spatty – git oot.'

We bypassed Spatty and carried on, across to Workington and down the coast to Ravenglass where the sandy bay glistened under the summer sun and tourists made sandcastles before visiting Muncaster, a castle constructed of more solid material altogether, or taking a trip on la'al Ratty, the narrow-gauge railway that once served the mines and now serves holidaymakers.

We too are heading for Muncaster since it is Patrick's home. He is the only friend we have who lives in a castle and he is every bit as unusual as his address. Born and brought up in Morayshire and educated at Eton and Oxford, Patrick really wanted to be Foreign Secretary but his tutor Tony Crosland got the job so Patrick became a sheep farmer instead. It was an obvious alternative to being Foreign Secretary. Especially if the land fires your spirit and your guiding principle is that man needs to live in harmony with nature.

Several rural organisations – the NFU, the CLA, the Deer Commission for Scotland and the Lake District Special Planning Board to name but a few – have harnessed Patrick's considerable talents down the years and his contribution and commitment to every single one has been incalculable. The staff at the Red Deer Commission, of which he was chairman for six years, honoured his retirement with a stag on the base of which was a plaque announcing: 'To Patrick – for annoying those who needed it.' He is very proud of that.

Yet Patrick remains a restless spirit, claiming to be just a tinker at heart, always on the road, making the countryman's voice heard above the urban din. What he doesn't tell you is that he fights like a savage on behalf of the misunderstood and under-represented and, as evidence of his singularity, numbered amongst his friends one eccentric who grew sweet peas up thistles and a neighbour who used to issue the utterly irresistible invitation: 'Come over, Duff, and let's swap lies.'

One astonishing man.

His witty and unconventional memoir *Those Blue Remembered Hills* chronicles the course of his extra-ordinary life and disproves the popular myth that the truth can only be told anonymously or posthumously and I need my copy signed – hence this trip. Besides, it is far too long since we saw each other. Like his memoir, lunch is peppered and spiced with some wonderfully indiscreet, desperately funny tales told as only he can – with a double dose of mischief.

A recent portrait adorns the front cover of Patrick's book and we are keen to see the original. It is hanging in the castle and the artist has captured his essence brilliantly: benign and thoughtful, with twinkling eyes backlit by naughtiness. People are teeming through the castle today and Patrick likes people. 'For every ten thousand delightful ones, there's one awful one. And it's nearly always a young man showing off to a girlfriend.'

People like Patrick too, even if they find him endlessly surprising. 'Hello,' he says to one visitor after another. 'Hello,' they respond. Then, quite suddenly, they realise that the face before them matches that on the painting hanging on the wall. They expect the conversation to end there but, Patrick being Patrick, it hardly ever does. He doesn't do vocal glad-handing: he chats because he finds humankind fascinating and visitors are invariably charmed by him. If people who live in castles are usually perceived as arrogant, humourless, distant and unfriendly, then Patrick pierces all those prejudices like so many balloons.

As a parting shot, he delivers his master stroke to an unsuspecting visitor: 'Thank you for coming and I hope you enjoy your day.' No doubt the little girl who learnt that she shared her name with a dog of Patrick's who tried to bite the Duke of Edinburgh enjoyed her day. I suspect her parents did too; inhaling Muncaster air is to breathe something restorative and clear.

*   *   *

It's showtime, big-time in August.

Just about every weekend there's an agricultural show in Cumbria, each one with its own particular, not to say peculiar, characteristics. Dufton showground, in the shadow of Dufton Pike, is a true local show. Many years ago, here, I saw a tiny girl on a tiny pony jumping a not-so-tiny wall; the wall was red and white and very possibly the same one responsible for Humpty Dumpty's catastrophic fragmentation. Pony and infant, though, flew it, if with a worrying degree of fresh air 'twixt backside and saddle, but the splayed red-ribboned plaits, jodhpured legs and pony all re-acquainted themselves on the landing side, and galloped off into the middle distance, a brakes failure clearly evident.

It's nearly always scorchio at Skelton, known locally as the biggest village show in the county and a marathon of organisation, co-operation and co-ordination. The only thing that really defeats them is precipitation – excess of, and that only happens once in a blue moon. You might be able to buy lychees at midnight in the metropolis, but at Skelton you can buy anything from a goodly slab of home-cured ham to a spanking new tractor.

One year, Carrs ran a Guess the Weight of the Sheep Competition, hardly a frivolous matter, considering that farmers need to be pretty accurate at assessing liveweight in order to arrive at the correct wormer dosage for their sheep. It was not a success: for a start, farmers' guesstimates were surprisingly hopeless, and worse, the sheep penned up for the competition (alias the Skelton

Three) legged it as soon as the pen gate was accidentally left open. An unseemly scramble across several fields and a memorable debacle involving a pick-up truck, one incandescent farmer and several red faces consigned this experiment to the file labelled 'Mistakes – Not For Deliberate Repetition'.

There were show classes for sheep too, of course. And for goats, dogs, horses, ponies, showjumpers, pets, beef and dairy cattle. Cattle are paraded, catwalk-style, in shining leather and brass headcollars. Mincing along elegantly does not come naturally to the average cow, and considerable training precedes any show-ring appearance. Tom Rawstron, a veteran – I am sure he will not mind me referring to him as that – of countless championship wins, explains. 'First of all, we tie them up for short periods and then we teach them to lead. It's a two-man job: the person at the front end hangs on for dear life and the one behind makes a noise when necessary by tapping their wellie with a stick.' I don't quite get this. Tom sighs, 'For acceleration, see? Then they learn to stand – head up, ears pricked, feet squarely spaced.' For the show ring, cows are selectively clipped; coarse blades for swathes of body, fine ones for head, udder, ears, tail and legs. Finally, show cows are pressure-washed – just on low volume, though. I was relieved to hear that.

Some, Tom says, are 'naturals' and love showing off, but others loathe being paraded about, and just shuffle along sulkily. They also need to be completely unflappable amidst the showground's unfamiliar sights, sounds and smells. Especially the smells: 'Well,' says

Tom, without any sense of irony, 'one of our show cows hated the smell of hamburgers.'

Frankly, if I were a cow, I'd think that was reasonable, wouldn't you?

Gyp and Tess would never make show dogs. Tess won't let me bath her, much less pressure-wash her. Bathing enrages her so that she bites me on the nose, but occasionally – and this is one such occasion – it is essential. Tess has been rolling in something interesting. Fox poo, I fear. Now the important thing to remember is that to a dog, fox poo is very like Chanel No. 5 is to a fragrant, stylish woman – an unmissable treat. In she went, neck first, then did a headstand with double twist and pike to maximise absorption of the noxious substance.

The dog we had when I was a kid had similar tastes in perfume and this led my mother to walk about the house with a handkerchief liberally doused in 4711, folded bandit-style, and tied behind her head. Mum could never afford posh pong and 4711 was very nearly as vile as the stuff whose aroma it was designed to override, but the bigger problem arose when she forgot she was wearing it and answered the door to an unsuspecting salesman. And then she had the nerve to wonder why I grew up as I did.

But that's a diversion. A delicate sniff of the ether in the orbit of Tess's personal space indicated that a bath was essential. And even if she bit my nose, it could not be as bad as the time a dear friend of mine ended up bathing a cat that had fallen in petrol. I think the story went something like, 'Come and help, please. Unless you come quickly, if anyone lights a fag, the cat goes up.'

You couldn't refuse, could you? Exploding cats are to be avoided at all costs.

I like livestock. I would just prefer not to share my house with several different species all at the same time.

The dogs are here by invitation. Only real loonies invite their ponies into their sitting rooms and the sheep are allowed in only one or two specific areas of the farmhouse: the freezer and the oven (and then only after they have been skinned and cured). Other than that, I expect my creatures to stay where they belong: outside.

Lately, I have been experiencing a little difficulty in maintaining this state of affairs.

It started with the bats. Everyone knows that bats are protected and that you cannot do anything to oust them forcibly. You can serve eviction notices, obviously, but I am sure I do not need to explain the futility of that strategy. Vespertine shadows bring out the bats. A healthy colony of pipistrelles lives next door in the chapel; the pipistrelles were here long before us and we have become accustomed to their presence on summer evenings. As the light fades, the birds gradually fall silent – the blackbird is usually last to turn in – and then the bats hold sway in the sky over our walled garden, swooping and looping in aeronautical manoeuvres of startling complexity.

During our first summer at Rowfoot, in 1982, my mother was visiting from the smoke and apart from a few trivial moans, mostly relating to damp, funereal quiet, pavements (lack of) – you know the kind of thing – she was getting to quite like it. We were sitting out in the garden late one evening and sure enough our little winged friends began circling overhead. There were quite a few of them: well, either that or the same one was going round again and again. Fast.

'Aaah,' said mother, 'look at the little birds.'

'That's not birds, mother. That's bats.'

For someone who claimed to be getting on a bit, she could shift, could my mother.

In those days, I thought the bats probably lived in the barn, but cottage guests report that they definitely live in the gable end of the chapel. So we do not just have bats, we have bats with a direct line to the Almighty and that really is scary.

So here's the gruesome tale of the Curious Incident of the Bat in the Night-time.

We always sleep with our bedroom window open, even when the Helm wind is blowing and in sub-zero temperatures: we're just funny like that.

You can guess what's coming next, can't you?

There was a rustling behind the curtain. The wind whistled again (I think it was something from *The Sound of Music* but I can't be sure). Something flew in. I disappeared under the duvet; my husband did what husbands do and got out of bed muttering 'Poor little thing'. I was muttering too but less sympathetically. I think my muttering went more along the lines of 'Get that bloody thing out of here'. And he did: he got it into the little bathroom, shut the door, opened the window and waited. The bat found its way out and peace was restored: this is what men are for.

The bat, I suspect, enjoyed this adventure so much he went and told all his little batty friends in the chapel about it. And one nipped round to try us out a week or so later. We went through the usual 'poor little thing' and 'get that bloody thing out of here' routine, but Bat Mark II was in an exploratory frame of mind and less keen on the bathroom thing. He (he could, of course, have been a she but I was in no mood for biological accuracy) zoomed out of the door, husband in hot pursuit. Bat and man swooped into the front bedroom. The newly decorated in gentle shades of gold with Rennie Mackintosh curtains front bedroom.

'If it poos on the carpet, it's dead,' I squealed, 'bat protection or no bat protection.'

Fortunately it didn't. The only good bat, in my view, is one in a zoo. And the next time I see one of those I shall exact my revenge on the species: I shall tap sharply on the glass and wake it up.

Since the bats, we've been invaded by more wildlife than you can shake a stick at: a toad, an earwig, a large beetle, a small bird and a vole that tumbled into the water meter shaft, condemning Tess to an entire week of expectant hovering at the head of the hole on Vole Patrol. We airlifted it to safety, naturally, but she didn't believe us.

The toad was only a baby and had been lying in wait outside the back door, taking advantage of G&T's late night pee routine and hopping, unseen, indoors. Proving conclusively that baby toads have little foresight, it snuggled down for a desiccating night in the heap of magazines under the coffee table. Happily, for the toad anyway, I elected to take a comic up to bed to read and pulled one, and it, from the heap. For a baby, it leapt impressively high and then disappeared under the couch. I emitted the squeal so thoroughly rehearsed in the bat routine and we formed a plan of action. I would fetch an antique fishing net from the dairy and Malcolm would manhandle the couch to dislodge the intruder. It all went fairly well, until the toad hopped through the holes of the fishing net (I told you it was a baby one, didn't I?) and sought refuge under the radiator cabinet, whence it peeked out playfully.

With commendable boy-scout imagination, Malcolm crept up on the toad with a glass pudding bowl and a piece of card and caught him. He then chucked him out of the back door and went to bed. We haven't seen him since: he's probably grown up by now with a whole swarm of dependent toadlets.

The same night a beetle crawled creepily – they know no other way – across the kitchen floor as I tucked the dogs up for the night. I killed it and shuddered my way upstairs only to be greeted by an earwig in the bath. Malcolm said that earwigs bit people. I said they didn't, surely. 'Those nippers on its bum are not there for decorative purposes,' said he who knows about such things, so I squashed that too. I am turning into a serial killer, I thought.

The next day, I paused at the top of the garden steps to see a blue tit fly head first into the upstairs window.

They do this sometimes: they see another bird, only it's not another bird, it's their own reflection, but being blue tits they are too slow in realising this to avert a head-on collision. Having nutted the double glazing, it was a bit stunned: well, you would be, wouldn't you? And it fell almost to earth, latching on to my sleeve on the way. And there it hung, as I screamed a blood-curdling scream that I really must explain to the neighbours, who had probably heard about my serial killing tendencies and assumed I was moving on to bigger challenges.

I wasn't. I just don't do feathers and I don't do flapping: witness the bat incidents. I am sticking to dogs, sheep and ponies in future. At least they are furry and don't flap, creep, sting, bite, hop or cling.

Our last visitors for the month include a brace of children. Children have always struck me as a dangerous hobby when there are so many nice, safe things you can do in your spare time, like paragliding or whitewater rafting. And babysitting? As far as I can see that is the best form of contraception known to man or woman, with the possible exception of winceyette pyjamas. The boy is appalled by dirt. He doesn't like mud, dog hair or dust but his sister is fascinated by the poppy heads on the plants in the front bed and takes to dismembering them with great facility, scattering clouds of tiny seeds at her feet. She watches, transfixed, as they fall to the ground and gasps in wonder at the notion that next year, with luck and a little rain, there will be a whole new generation of poppies growing. Her brother has a sly pick of his nose when he thinks I am not looking. I am.

Did I say he didn't like dirt? He must just be selective about the kind of dirt he abhors and bogeys don't count.

Later in the evening I come across another child in the field behind the house. This is not a home-grown child or a related one but a visiting, articulated one. It is running full tilt towards the sheep. This child must be deranged. Why else would it give every impression of picking a quarrel with a Manx tup with four horns? It is shouting, too. Screeching. The tup stands his ground and begins to look faintly murderous.

The tup paws the ground and I find myself turning into Joyce Grenfell.

'Don't do that, he won't like it,' I announce, my voice several octaves above its normal range.

The mutinous child scowls at me but reluctantly gives in, backs off and withdraws. For a terrible nano-second, I fear that the tup may chase it, impale it on his horns and practise his caber-tossing skills.

I am very glad that I own sheep not children.

Just as I am realising that stocks of Big Frank are running low – it is quite amazing how quickly it is possible to consume so much of something so tasty – we receive a slightly frantic email, headlined 'Piggin' Disaster', from Justine and Paul. Usually I find that pig farming by proxy beats the real thing hands down, having all the advantages of involvement and satisfaction without the need to deal with swill, dung and movement licences, and, naturally, I take my role as pig consultant very seriously. Jus and Paul have a problem, or rather nine problems, six dead and three alive. Their prized Large Black sow Barbara farrowed two weeks before her due date, producing a litter of very tiddly piggies. Only a reckless gambler would have bet on any of them surviving, so Jus and Paul have done well to salvage three from the wreckage. The trouble is, when they are your pigs, it doesn't feel like that.

Barbara is toughing it out with the remaining three and Paul says that she is proving to be an excellent mother. If you have never kept pigs and your association with them is informed only by children's storybooks in which they squeal, snort, run mischievously about farmyards and do unlikely things with tractor wheels, I have to tell you, gentle reader, that this is a euphemism. What it really means is that Paul needs to wear full body armour at feeding time because motherly old Barbara is pawing the ground, swaying her head in a homicidal fashion and threatening to kill anyone who comes anywhere near her babies. And if that came to pass, Paul would not need to bother about feed for Barbara because he would be it. I dare say she'd leave a few crunchy bits, but dental records would indubitably be required for positive identification.

A week or so later, feelings are less raw and Jus and Paul ring to say they are bringing Donny. Donny, Paul assures me, loaded with considerably greater ease this morning now that he is packed in plastic bags than he had done last week when he still had full mobility and retained the capacity of independent decision making.

Donny comes to us, but his brothers Jay and Jimmy and his sister Marie, collectively known as the Osmonds, are back in Lancashire and quite possibly rehearsing a reprise of 'Crazy Horses' as I write. Their playmates are a family of Large Blacks called the Jacksons, one of whom, Michael, has poor skin pigmentation. So, you see, it is true that life imitates art.

While Donny reposes quiescently in the boot we hear that Paul has been busy constructing arks for his pigs. He didn't like the price of the bought sort, you see. Pig arks, Paul informs us in scandalised tones, cost £300 plus transport from Newmarket. He did think of asking a racehorse trainer to stow them in the back, behind the horses, under the solarium and in front of the hay on his way back up the M6 but it was a scheme that appealed to few, so Paul followed Noah's example and built his own. He designed them to be cosy, efficient and most importantly cheap, so utilisation of some recycled materials became an imperative. There followed a long and sorry saga, revolving round the collection of a quantity of corrugated metal and a lorry stuck in heavy plough. Paul's first plea for assistance fell on stony ground. 'The bastard just looked at me and said no. So I looked him in the eye and said, "I hope your willy falls off."'

No serious harm intended, then. I think he was quite restrained in such trying circumstances; most normal people would have added 'but not until it has developed incurable and extremely virulent gangrene' . . .

Someone else took pity on him, though, and the arks, you will be pleased to hear, are a great success.

September

Tomatoes hang heavily on the vines in the greenhouse now. My favourite supper through the cropping months is grilled tomatoes on toasted homemade granary bread. I am very cheap to keep. Elsewhere, the harvest ought to be well and truly home. If it's late, you'll be sure to hear someone in the back pew by the door muttering that he's not giving any thanks for summat he ain't got yit. The Cumbrian harvest is predominantly live as commercial lambs approach finishing and the important pedigree and breeding sheep sales of the year get under way.

I HAVE ASKED ERIC to look in the barn for owl pellets (for the uninitiated this is owl sick or poo, depending on how you look at it: the regurgitated waste bits of the tasty morsels they've dined on). I have had a message from Jenny at the World Owl Trust asking if I could possibly a) find out if the local barn owls have reared any babies this year (sadly not) and b) send some pellets to Muncaster for analysis. So I've asked Eric to grub about and see what remnants of his ghostly lodgers he can find. He is also supposed to be quoting for the work to the Men's Bothy and delivering some hay, so really he shouldn't have time, but he's a good-hearted soul, Eric. And thoughtful with it – witness the exchange we had.

'I'll have a look on top of the big bales in the old barn.'

'Great.'

'And I'll collect some.'

'Good. Could you put them in an envelope, do you think, and shove them through the door?'

'Yes, no problem.' Small pause. 'I'll seal the envelope up so that the dogs don't get them.'

See? Thoughtful.

The barn owls live amongst the beams and rafters of the old barn at Townhead and swoop down here every now and then to visit. I hear them at night; a chilling sound it is, not the gentle mournfulness of the tawny owl but a different, more troubled cry altogether. Troubled they are, our beautiful native owls, as their habitat is eroded by modern farming methods and the character residences they loved, with nooks, crannies, corners and handy perches sited on every wonky wall, are replaced by those hideous green hangars beloved of modern agribusiness.

I would love a barn owl to come and nest in a Rowfoot barn. I've put out signs (invisible to the human eye and to be read only by educated owls) saying Owls Welcome, I've put an owl box in the wood (though there's probably a magpie in it), but thus far my efforts have been to no avail. I live in hope and until then I shall have to content myself with the gracious tawny who sits on the gable end of the house on still nights and looks down on me with a quizzical expression. She's beautiful.

What with modern farming and its soulmate modern traffic, it's small wonder that the owl population is struggling. Anyone who thinks that we don't get much traffic noise in the country should think again. We don't get buses, of course, except the school ones each morning in term time, one public midi-sized one on Fridays and another on Saturdays, but we do get tractors the size of Chieftain tanks with spiky accessories on the front and a couple of trailers on the back. We get milk lorries weighing as much as oil tankers and livestock wagons so multi-tiered they rattle up the angels at night. If any combination of heavy freight meets on the corner by our house, everything has to breathe in first and find reverse gear next.

There used to be two particularly lethal motorists who lived in Ainstable; I shall not name either for fear of litigation but I was always confident that if they were heading in opposite directions and met on our corner both would go on their respective merry ways completely unscathed. How would they manage this? Simple: both of them would be on the wrong side of the road, so they would pass each other in absolute safety.

Chickens don't venture out into the roads much at all these days so I was quite surprised when one of those email round robins asked recently, 'Why did the chicken cross the road?' It went on to postulate that the Iraqi

Head of Information might answer that there is no chicken, never was a chicken and even if there had been a chicken at any time it wasn't on the highway. Me, I think road crossing is best left to zebras.

Due to my deeply entrenched feather phobia poultry rarely get any kind of look-in in my scribblings but complaints have been pouring in about this from the Feathered Action and Representation Team (another of those organisations best not reduced to an acronym, I'm sure you will agree), so the chicken gets a walk-on-and-don't-get-squashed part this time.

That's it, chickens; your moment is over.

It's still important to explore the matter of road crossing more closely. Did you know that some 270,000 creatures (and that's not counting the fleas worn by the hedgehogs) lose their lives in road traffic accidents each year? It's a startling statistic.

If you are a fox, crossing the road is a potentially hazardous activity if that wicked pixie Jeremy Clarkson is scorching about in a turbo-charged babemobile because he, given the slightest chance, will make you into mince. He said as much on *Have I Got News For You*: what is more and what is worse, he said he did it deliberately. That's the Mr Clarkson we all know and love.

Foxes, like anything else left for dead on the public highway, become the 'property' – I love that – of the county council: some of the latter's less desirable assets, certainly. In an inventory of Cumbria County Council's possessions, foxes, badgers, hedgehogs and toads (deceased) must come somewhere between educational establishments and social services, but that is who they belong to. So if you find one and want it moved you now know who to call . . . I'm sure they'll be delighted to help.

While we are on the ticklish subject of foxes, all that fuss about banning hunting could have been avoided altogether with just a bit of common sense. There's little dispute that the fox population needs controlling, but as it is too much to expect Jeremy Clarkson to carry out the task with only a single combustion engine at his disposal at any one time, the challenge of finding the right, humane and acceptable means remains. In the light of statistics showing that traffic accounts for a great many more foxy casualties than the most effective of hunts, had the revered and reviled of Westminster simply commissioned the construction of a few new motorways, bypasses and flyovers they could have killed two birds (and several foxes) with one stone. It would have saved an awful lot of trouble.

Round here there's a novel highways policy: it's a creeping road widening scheme and it happens like this. Those ever larger tractors, wagons and lorries, all dismal testaments to that fiscal imperative known as 'economy of scale', pulverise damp grass verges into an unrecognisable pulpy gumbo from which they never recover or regenerate. Then other vehicles drive in the sludge, and when a trough develops 'twixt verge and existing surface an operative in a custard-coloured mac comes round and slings a load of tarmac into it and instantly the road is wider than it was before. There you have it: highway robbery, twenty-first-century style.

Only when the walls start falling into the roads will anyone begin to take any notice and by then, of course, it will be far too late. The stoats that live in the hedgerows will have packed their belongings into their micro Louis Vuittons and scarpered, along with weasels, voles, field mice and the cream-coloured hamster that escaped from the local primary school in 1999 and has been posing as a tailless albino mouse ever since. The wild orchids that surprise us with their astonishing fortitude and pop up each year will be pulped like so many unsold books and the last refuge of the vulnerable pedestrian will have been irretrievably mangled.

Still, the number crunchers are not worried and that's the main thing these days.

Motorists are not required by law to report accidents involving foxes and other small mammals but if they clout a dog, horse, ass, mule (or any other calamitous equine mating, presumably), pig, sheep, cow or goat then there is a duty to let the constabulary in on the details, however gory.

Running over a dog is undesirable but possible unless it is something like a St Bernard or a Rhodesian Ridgeback, in which instances running 'over' becomes unlikelier than running 'into'.

It gets tricksier still if you run into a horse. People greet me when I am out riding with the hand they are not using to support the mobile phone, raising the spare very briefly from where it rests tenderly on the steering wheel, a delicate implement that they don't want to wear out through overuse. They wave. And this is the best bit: they expect me to wave back. I think not. I like to keep both hands on the reins and my brain on the job. Funny, but that's just the way I am. So take my advice and avoid hitting horses – hoofprints on the Daimler are not a fashion accessory. And hand to hand combat with riders is best passed up on too.

I imagine that bovines are specifically targeted by yachties during Cowes week, but in the unfortunate event of collision between beast and vehicle in, say, the open pampas of Crouch End, the bull bars on the 4WD minimise the catastrophic effects of impact. On the narrow bendy lanes of rural Cumbria, though, this is a real consideration. To the irritation of other road users, I drive as if I might meet a flock of sheep round the next bend, mainly because I often do. You do not want to hit a flock of sheep, and goats should be avoided at all costs. Not specifically in the car-collision sense, either; just keep away from them. No human is a match for a goat. If a car hits a goat it may be an accident, but it may just be that the goat was in a confrontational frame of mind anyway. They often are, you see.

On meeting a goat, a car driver can expect the following sequence of events to unfold. The goat will lower its head and swing it menacingly, butt the front bumper, and then dance on the bonnet leaving nasty pock marks just below the windscreen wipers before leaping, capriciously of course, on to the roof. Here, the goat will stand on its hind legs and deliver a critically acclaimed rendition of the ditty so sweetly warbled by Julie Andrews in *The Sound of Music* about lonely goat turds, lay-holla-lay-holla-lay-he-hoooo, before coughing and dismounting from the car. As a parting shot, it will kick the rear number plate and leave, never to be seen or, thankfully some might say, heard again.

Take that as a warning.

A final thought on the original theme of jaywalking poultry: is it me, or is spatchcocked chicken really recycled roadkill?

Mind how you go . . .

Feet up time and I'm rummaging in the local paper again. Now the journalists are suggesting that we are in the midst of a rural crime wave. Speeding pensioners, sheep rustlers, rules on red diesel being flagrantly abused and machinery being nicked from outlying steadings: it is not exactly murder and mayhem but our understanding of the word 'crook' is beginning to extend beyond sticky things made from hazel twigs.

I hope that the speeding pensioners take my observations on road safety to heart and my second group of miscreants might be put off their stride if they ever attempted to rustle Manxes, but crime is on the increase, even if Nick Ross assures us at the end of *Crimewatch*, that vicarious orgy of rape, pillage, arson, larceny, fraud and murder, that it isn't. He'll be telling us next that Mr Hitler was a good bloke who didn't mean any harm and *Walking With Dinosaurs* is reality TV.

Nick-Nick cites 'statistics' as evidence. I remain unconvinced. Until very recently, in Cumbria we regarded locking doors as an urban custom, but now 70 per cent of us are frightened of being burgled. Mind you, people living in southeast England might be forgiven for fearing muggings, random vandalism and being murdered, not to mention the possibility of getting caught up in a terrorist attack, so maybe we Cumbrians are just a bunch of old scaredycats. Somehow, I fancy that the days of scrumping, sheep rustling and the licensee of the Ferret and Flymo's passing off of VAT 69 as Jerez's finest when every schoolboy knows it is really the Pope's telephone number are no more than fond and distant memories for rural constables.

Speaking of policemen, out here in the sticks you are about as likely to see one as you are to tread in a heap of rocking horse do-dos. Years since, there was a Hofficer of t' Law at Kirkoswald, about two miles up the road and over the hill, and another at Hesket – same distance, opposite direction. So two within, say, six miles of each other. There wasn't huge scope for much misbehaving. Nowadays, there is one officer, in a car, to cover over four hundred square miles. You may contend that this area is populated by more sheep than people and sheep are responsible for few petty thefts and still fewer muggings, but you see the point. But the really clever thing is that it took a whole raft of urban experts to plunge us into this absurd situation, opening the floodgates, field gates and farm gates to all sorts of nefarious activities from drug dealing to failing to fill out Animal Movement Records.

And pity any officer who chances upon some miscreant, because it is his bounden duty to recite that stupid modern caution starting 'You do not have to say anything. But if you do not mention now something which you later use in your defence, the court . . .' and going on for long enough for him to take a refreshment break mid-spiel. He could use the Alternative Cumbrian Vernacular version – 'You needn't say owt. But if thoo hes owt to say, thoo'd best say it noo . . .' – but sadly it's not legal. This is a great pity, as it would save valuable man-hours.

Farmers take few chances these days, locking up anything operational with the possible exceptions of the dog and the spouse, both of which are security post-coded instead. And we're all signed up to Farmwatch, of course, to keep everyone informed of what is being lifted from whom. Not that long since, a few gates went walkabout but their loss was not reported for days on end. Gates, I hear you say. Surely you'd have noticed if gates had gone missing? Wouldn't sheep and cattle be cluttering up the highways? Well, yes, I thought so too, but apparently not. Even though these were not just any old gates but particularly tastefully finished ones, possibly with gold finials or solid silver snecks – at least that's what their estimated insurance value suggested . . .

Some thefts are opportunistic. Others are cleverly orchestrated, planned crimes. Take quad bikes. Some villains did. Since this thieving was almost certainly to order, they had to arrange their despatch. And how did they do this? They bought, stole or appropriated some straw, a lorry and a flatbed trailer; then, ever so care-fully, they lined the quad bikes up on the flatbed and boxed them in with straw bales and light timbers, balancing more bales to make a dual purpose ceiling cum floor on the cross beams – I expect there was some baler twine involved somewhere; there usually is – then on with more quads and lash the whole jolly lot down and set sail for Ireland. How neat is that?

For ineptitude and sheer bad luck, though, you would have to go a long way to match this little tale. Are you sitting comfortably? Then I'll begin.

It was a dark and stormy night (dark and stormy nights being an essential prerequisite for tales of dastardly deeds). At 5.34 a.m. the phone in a farmhouse not a squillion miles from here trilled. The phonee trotted downstairs, worried. Phone calls at such an ungodly hour herald events dire and disastrous. Perhaps an elderly relative has been hospitalised, arrested or shot. Or Auntie Violet has failed to notice that the little arrow on the petrol gauge strayed into the red zone somewhere about Stoke Poges thirty-six hours ago. Maybe friends fuelled by 4X in the Sydney suburbs have catastrophically miscalculated the time difference. She picked up the receiver. 'Do you own an Ifor Williams trailer?' Whatever was expected, someone doing market research on horse transport purchases was not it.

'Yes, it's in the yard.'

'Oh no it's not.' If smirks were audible, this one would have had an echo. 'It's fifty yards down the hill.'

Our brace of varmints had hauled the trailer across the road to the waiting getaway vehicle, a Vauxhall Cavalier: brigands on a budget, clearly. It was then that they began to hit serious snags, a security bolt through the hitch having completely escaped their notice. Undeterred, and without a thought for their personal safety, they decided to rope the trailer to the car. At that moment, a police car happened by. This was in the days when policemen went out in pursuit of villains instead of obsessing about targets and figures and cost-effectiveness. The officers introduced themselves. It is unlikely that the exchange included the words 'Evenin' all' for the thieves abandoned the trailer fifthwith. That's just a little faster than forthwith. One jumped into the car, crashing it at the first bend. The other legged it across the fields. Then it started to snow. As the saying goes, there is no such thing as the wrong weather, just the wrong clothing, and happily this quite staggeringly inept crook was wearing exactly the wrong clothing. He was found huddled in a phone box several hours later, phoning his mum for a lift home.

Not a good night's work, one way and another.

The creepiest thing about all this was not that they nicked the trailer, not that they both looked unspeakable little thugs, not that one of them had touchingly secured an apprenticeship as a plasterer with his uncle's firm (I made a mental note of the firm so that I could avoid it in future; I was not persuaded that he was going so straight in future that he would forgo the opportunity to case the joint), nor even that his girlfriend was pregnant – though I fervently hope he had a bag over his head when he did the deed because he was quite the ugliest young man I had seen in many a moon. No, it was this: that these two villains had been prowling about my yard, sizing up which of my things they would steal, while I was sleeping soundly in my bed a matter of yards away. That was what upscuttled me most of all.

On the whole, though, orgies of criminal excess are not commonplace in rural areas. Local crime is either of the organic home-grown variety involving Animal Movement Records oversights, driving offences and drunks, or the imported sort, involving thieves, robbers and vengeful ex-spouses with shotguns.

Sometimes, they collide, and when they do there's usually a milestone birthday in the mix. Naturally, what follows is a tale of fiction but not beyond the realms of possibility.

Let's start with some fiftieth birthday celebrations. There's the usual bedsheet daubed 'Fred's 50 Today!' in primary colours and strung up alongside the A6 without planning permission. The party follows and everyone gets very, very drunk and then has to crank their brain into gear to deal with the business of getting home. The bright ones doss down in a barn, the others don't. One or two take a circuitous route home in a JCB along the byways and tracks, gathering some very nasty scratches on their paintwork as they go. Another leaves early having drunk an entire bottle of Southern Comfort and several pints of beer. He claims to have supped the Southern Comfort for a bet; the beer was just for the hell of it. The balance of his mind being a little out of kilter, he decides that he is perfectly capable of walking on water (happily, the party is a few miles from the river), unassisted flight (thankfully there are no high-rise buildings in the immediate vicinity) and, more prosaically but just as dangerously, driving home. He did, or at least he started to, and that's where we'll leave him for a moment, if that's OK. Don't worry, he will come to no harm.

Not far away, a gamekeeper has been having a little trouble with poachers. He calls the local constabulary to alert them to the presence of crims in the countryside but the constabulary are busy elsewhere, dealing with drunks, skunks and drug dealers in Carlisle. The gamekeeper is quite keen on natural justice too and this, he thinks – without the aid of Southern Comfort or beer – is the perfect opportunity to dispense some. He and two pals leap into a pick-up and set out in hot pursuit of the poachers, successfully apprehending one but missing out on a further brace who escape unscathed.

What to do with the captive? Make an example of him, obviously. The good guys decide to teach the baddy a lesson he will be unlikely to forget in a hurry so they strip him naked, handcuff him (yes, handcuff: where did they get them? Best not ask . . .) to the dog cage in the back of the pick-up and set sail with him. Our cold poached felon tries to jump out – God knows why – and ends up getting his important little places uncomfortably gravelled.

The police by now have extricated themselves from dialogue with the urban lawbreakers and hasten to the scene. On their way, they come across a vehicle weaving across the Queen's highway in a most disconcerting manner so they stop it and breathalyse the driver. And then they nick him.

Next – and they are all agreed that this is turning into a most productive night's work – they find an in-the-buff poacher crouching in a pub car park, his backside cruelly gravelled.

Some time later, the villains of the piece are brought to justice and the good live happily ever after. It's worth adding that any villains in this part of the world tend to be much less bothered about due legal processes than they are about the reactions of their mothers, who send them all to bed early for a week without any supper.

And the moral of that little tale is that the crooks shepherds lean on and wave at their dogs are not the only ones in the countryside.

Elaborate burglar systems are great but here at Rowfoot we favour rather more bespoke deterrents. They are called Mickey and Blossom. They are big, they are hairy, their hooves are shod with one eighth of the national output of British steel and they kick with the kind of unerring accuracy that Messrs Becks and Owen fantasise about. Mickey and Blossom's footwear might be costly, but razor wire, broken-glass-topped walls, electronic security gates and digging a moat might turn out to be even more expensive. Oh, and they don't like their beauty sleep being disturbed by strangers one little bit.

What can be done to protect isolated rural property from marauding thieves? Shod hooves apart, I've given this a bit of thought, mainly when we have just locked and left and before I start on the really irrational stuff, like wondering if I left the iron on and the oven off or the other way round, or both, or neither. I am not the only one dwelling on the matter, as apparently police in Derbyshire are in cahoots with a garden centre promoting a range of particularly prickly plants. Just for crime prevention purposes, you understand, not for commercial gain. The Derbyshire police remind us, very thoughtfully, that brambles, roses and berberis provide a barrier and a deterrent to intruders. Quite, but this is the bit I really love: sharp thorns, they say, can be used to gather DNA evidence. So, berberis – or, as I prefer to think of it, environmentally friendly razor wire – keeps burglars in shorts at bay, whilst containing your children and dogs very prettily. I bet you never thought of that.

Gravel serves as an early warning system, of course; a pond may act not just as a mediaeval moat if you site it appropriately but also as a leisure centre for your geese. And believe me, geese are a better deterrent to intruders, not to mention close personal friends, distant relatives and nomadic worm reps, than anything else you will find. The tricksy bit with geese is persuading them to yield up the mouthful of tasty fresh buttock from which the nice forensic man can extract the wretched DNA. They are assuredly more use than Man's Best Friend, chiefly because if yours are anything like G&T, they are somewhat indiscriminating about how far, exactly, they extend the paw of friendship. They are easily bought. A Bonio or a bar of Curly Wurly and it would, I fear, be 'Come in, mate, what's mine's yours . . .'

You could fix a dummy burglar alarm box to the outside of your house, indicating that you might have something inside worth nicking to start with. You could install advanced closed circuit television and spy on those misguided enough to think it was they who were spying on you. You could leave a pair of sturdy boots on your doorstep or in the porch in the hope that the man with the arrows on his outfit will think the boots' owner is a resident; you could shout 'Bye for now' to no one at all (though if your neighbours notice this they could become unnecessarily worried for your sanity); you could buy a wreck and leave it parked in the yard as a professional decoy.

Failing all that, do what I did: marry a copper.

Crime has always been writ large for me ever since the incident of Robin and the Sherbet Pips. Robin rather liked me when I was seven. Early evidence of a discerning man? Probably not. Probably more to do with the fact that I liked his bantams, rode a bike recklessly and willingly played in goal when no-one else wanted to. I reciprocated his affection; after all, he was the only one in the class with a noticeable absence of bogey-candles. Ours was a match made in heaven and that is why he bought me a bag of sherbet pips. Being none too dexterous we dropped the whole lot on the floor during Mr Green's history lesson one Friday afternoon. Mr Green saw red. He shut Robin in a cupboard (damaging him for life, no doubt) and made me clean his cricket whites. By the time I went home after school, I had splashes of white decorating my face, uniform and shoes. I prayed my dad would not notice.

'What the devil have you been doing?'

Stupidly, I explained about Robin, the crime of the sherbet pips and the punishment meted out. My dad sided, predictably enough, with Mr Green and smacked the back of my legs.

I have harboured serious doubts about justice ever since.

Before you think that this level of criminality is enough to amuse hobby-bobbies, specials and paid officers alike, just think what lies round the corner for the rural constabularies now that foxhunting is outlawed. People who are likely to have done nothing more reckless than park on a crack in the tarmac in the past will soon make up the greatest number of mounted criminals since the heyday of the highwaymen some 250 years ago: that's real progress.

Proscribing an activity rarely silences debate about it and so it is with hunting. Opinions are firmly entrenched and it has all become decidedly ugly. Yet there's something very odd about the arguments. No one seems to question that, sometimes at least, foxes are a pest and that sometimes pests need killing. So it seems to be more a question of how foxes should be despatched and whether, in a civilised society, it is OK to own up to having enjoyed chasing something to its death. Of course, no-one does own up, and over there on the moral high ground the antis are shouting 'cruelty' while the pros are shouting 'civil liberties'. There is a shoal of red herrings in the shallows: class war, employment, snobbery, tradition, envy and a few other prejudices besides.

Bizarrely, nearly half a million people took to the streets to protest against a ban on hunting while polls show that more than 70 per cent of the great British public favour just such a ban.

So far, so inexplicable.

And that's the trouble with the hunting question: there are no universal truths. As a means of controlling the foxes, hunting may be less efficient than roadkill but at least the end is sure: the fox either escapes unscathed or winds up very, very dead indeed. Some hunts perform a valuable community service, collecting carcasses from farms; others don't. Some hunts diligently seek permission to hunt across private land; others go without permission before or thank you afterwards. Some people will tell you that no foxes are reared for the express purpose of hunting them – and others will assure you that this is a vile calumny. Some claim that only the weak and the old are caught but some old and weak manage to hobble far and fast enough to generate reports of terrific runs for many miles across country that challenges horse and rider. And there is no point in hoping that drag-hunting will catch on because it won't. While farmers perceive a trade-off of sorts under the current system, reasoning that one dead fox equates to several live lambs, they are none too sure that they want their fields ploughed up a couple of times a year just so that drag-hunters can have a bit of fun riding cross country.

In the uplands things are very different. The invocation of traditions of venery and moral certitudes are all pretty superfluous up there: it's pest control, pure and simple.

See what I mean: no universal truths.

The fox remains a pest who kills for blood-lust as much as subsistence, and while we humans are troubled by the concept of 'fairness' in relation to his fate, the Act that outlaws hunting fails to outlaw the arguably more iniquitous practices of snaring, gassing and poisoning. If it's all about animal welfare, that's a little odd to say the least.

The fantastically eccentric lady Master of Foxhounds I worked for in the south of England in my gap year spent countless hours 'summer hunting': visiting farmers and landowners to obtain permission to cross their land during the forthcoming season. I drove the beaten up old Land Rover on such occasions, ferrying her from one farm to another as she stroked egos and drank gallons of 'filthy tea'. She much preferred whisky. Once, she came out of a particularly scruffy farmhouse wearing a beatific smile and waving to the farmer as she tipped the terriers off the front seat. 'Bye-bye, thank you *so* much. The hunt greatly appreciates your continued support . . . Odious little man,' she muttered resignedly as she lit up a Woodbine and inhaled deeply, 'but I need him on side.' She loathed it, all that public relations and talking and smiling, but she knew it was pivotal to the continuation of her winter fun and bore the yoke of peacemaker stoically.

She insisted on good manners from field and followers, firing volleys of invective at anyone parking inconsiderately, leaving horsebox ramps down, or failing to clean up droppings before they went home. Perceptive and pragmatic, she well understood that transgressions like these had a definite impact on the

everyday lives of ordinary people living in a changing
countryside even if none of them was remotely connected with the morality of hunting foxes. She was way
ahead of her time.

Yet, while hunting has aroused wrath and emotion
aplenty, chickens are still being battery reared and,
worse, the poor old pig is still being factory farmed
before being packed in blue polystyrene and sold on
supermarket shelves. Factory-farmed pigs live not just
a fraction of one day as a persecuted quarry but the
whole of their miserable lives in the most unrewarding surroundings imaginable, but no-one takes to the
streets, puts up posters or harangues politicians about
their fate.

We should honour the noble porcine, not mistreat
him, if only because he is an accurate forecaster of
fashion; pigs, may I remind you, were into body piercing
long before humans. Evidence of this is clear from
Edward Lear's writing: 'And there in a wood a Piggy-wig
stood, With a ring at the end of his nose . . .'

Fighting on behalf of the fox. Overlooking the pig.
What a curiously selective lot we humans are. And that's
before we even get started on the Animal Health bill: in
a satisfying little twist, among the suggestions in the
consultation document is one that all animals should be
entitled to act out their 'natural behavioural pattern'.
Nothing wrong with that. I'm all for it. But where will it
leave the hound, who is, after all, doing no more than
enacting his natural behavioural pattern in pursuit of Mr
Fox?

We're doomed, all doomed, I tell you.

October

We like cruising. We like DIY. One year, very misguidedly, we put the two together. I realised the full extent of my mistake when I heard the voice of a demented fishwife shrieking at my husband, who had just fallen off the boat and gashed his leg horribly, 'God pays debts but never in money.' The voice was mine. We let someone else do the driving now. It's better that way.

OCTOBER: SEASON OF mists and mellow fruitfulness, Keats said. Even taking into account that Keats's opium habit might have impaired his meteorological judgement, it is mellow, and this year the wet summer has meant that everything is still green, still lush, still rich. There is so much grass that farmers are taking extra cuts of silage and stock are still feasting on good pasture. They are eating their greens and that's always good. Ultimately, I suppose this means that if you eat beef, you too are eating your greens, even if they are delivered by a slightly devious route. Who knows, a T-bone steak might inadvertently furnish you with the entire daily allowance of vegetable products recommended by Her Majesty's government.

The first day of October always brings back memories as it is our wedding anniversary. It is twenty-nine years this year: I thought I had better write that out in full as it looks slightly less alarming in words than in figures. Neither of us feels old enough to have been married that long, though Malcolm jokingly (at least, I think he is joking; there'll be trouble if he isn't) says that he could have killed four times and still been a free man.

Mickey is coming home.

Sarah has decided that he is unlikely to set the eventing world alight and she has decided to send him home to his old mum and look for something that possesses the same ambition and capacity for self-improvement as she does. He is capable (enough), jumps well (enough), but he is undoubtedly a horse for whom the phrase 'can't be arsed' could have been invented. He lacks the killer instinct that made his mother, Kareima, such a formidable competitor. At the start, as they counted us down, I'd be hanging on to her, trying to prevent her from jumping the starter's gun. Mickey, on the other hand, waits for five, four, three, two, one, *go* and pauses for a second, as if to say, 'Oh, my turn now, is it? Jolly good, chaps, see you all shortly,' before pinkling off in his own time.

Perhaps he just fancies early retirement. At the age of seven, it is very early retirement indeed, but there is plenty of time for him to try his hoof at something else. He has the looks to show, and presence by the shedload; he is the only horse I have ever seen who looks in the mirrors set around an indoor school to make sure they are reflecting his good side. He has sufficiently extravagant paces for dressage and the breeding for long distance, so finding an alternative career for him should not be too hard. Recklessly, I have said I will keep him ticking over; rather like humans, it will be easier to find him a new situation whilst he is gainfully employed, at least part time, than from a standing start.

The options are as follows: sell him on the open market. Can't do that. How do I know that he won't end up in the hands of the horsy equivalent of the white slave traffickers? My mother used to worry about them, chiefly I suspect because she wondered what the going rate for a small, annoying red-haired girl might be and what on earth she would do with two camels in Barnes. So, as he is really my baby, it follows that I ought to worry about them too.

Option two: find someone else to have him, on loan. Who? What will they do with him? Will I trust them as I have trusted Sarah? Will they love him enough? Too difficult. Too many sleepless nights lie that way, I fear. Incipient madness cannot be completely ruled out.

Option three: keep him at home, for now anyway. All he is going to cost, I explained to Malcolm, is his food and his shoes. And time, of course . . .

I will give you three guesses as to which choice I've made. And your first two goes are invalid because it is just too easy.

One horse: fifteen jobs.

'His saddle's coming back with him,' I explain to

Malcolm who is wearing a benign, if not actually vacant, expression of total surrender to the inevitable. So we have to find somewhere to put a saddle rack in the dairy. For 'dairy' also read 'tack room' and 'all purpose glory hole'. The dairy is the sort of place where you put everything you cannot immediately find the right home for. It is a tip most of the time but at least you can shut the door on it.

Malcolm stands in the dairy and looks round hopefully, expecting a space to materialise from nowhere: there is none. 'That cupboard will have to come down. The saddle rack can go in its place and the cupboard can go over there.' Another gesture, vaguely in the direction of the bread bin. 'That will have to move and so will the spice rack . . .'

It starts to sound rather like that song they used to play on Children's Hour about the Day the Gas Man Came to Call. Half an hour later, we have decanted the Hoover (broken, again, in spite of a new motor's being fixed only in April – how I wish I had bought a Dyson), several horse rugs, two saddles, the bottles for recycling, a heap of newspapers, the Kenwood Chef and a length of plastic-covered electric wire that I use for poking down the tube of the self-defrosting fridge to clear the gunk that grows in it, two Calor gas cylinders we had forgotten about, my three-legged milking stool, its seat polished to a sheen by my bottom, and a dead potato (circa 1987) into the kitchen. Out in the workshop Malcolm constructs a shelf to go atop the sandstone one encircling the dairy and in the dairy I wipe every horizontal and vertical surface, removing cobwebs and generally tidying up. I don't bother dusting the potato, though. That seems a bit de trop.

We might as well decorate, I say.

Armed with a five-litre pot of masonry paint, bit by bit I douse the walls, and gradually the contents return to their rightful place, not to be confused with the place whence they came. It sparkles now. And it houses another saddle rack.

Mickey has grown so much in the past year that while you can get him out of the stable wearing his saddle as it is slightly downhill, it is no longer possible to lead him slightly uphill back so I have to remember to unsaddle him outside. Also, due to his late growth spurt – possibly due to his Irish Draught blood, possibly due to the Arab (blue) stuff, I know not which – he has to buckle his knees, twist his head and rearrange his neck to get his face over the stable door. This is a highly unsatisfactory state of affairs, so Malcolm agrees to fit a slip rail so that Mick can watch the world go by in a degree of comfort.

We trail off to Relphs and buy slip rail brackets, to WCF for a saddle rack, to the timber yard for a length of four by three, to the saddlers to buy the bridle and the bit, and to Carrs for some wormer, fancy body-building feed, hoof oil and mock-creosote for the slip rail lest he tries to eat it and poisons himself. Next we visit the farm shop in Penrith to purchase an industrial quantity of carrots. I send off for a winter rug (rip stop breathable outer, quilted polycotton inner, leg vents, high neck – it costs more than my own last winter coat but it lacks the nifty fur collar. Still, you can't have everything). I have paid the insurance, phoned the blacksmith and bought in some new season hay. I must get some new brushing boots too.

I said it wouldn't cost much to have him home.

Lionel and Peggy arrive for a week at the cottage. Lionel has read *The Funny Farm* and thought that Rowfoot would be just the place for a late summer holiday, as the leaves on the trees started to turn and a milky bloom of light frost lingered on late flowering roses. I don't think they were disappointed.

Lionel and Peggy have a small herd of pedigree Highland cattle. Knowing that they were abandoning ship for a week, they left their precious cattle in the care of a fellow enthusiast, also taking the wise precaution of having the vet over to check old Kirsteen who is in calf but not due yet. Not for at least a month, said the vet, having investigated Kirsteen's innards. So they came away from Hampshire, clear of conscience and con-fident that their return home would be well in advance of Kirsteen's calving.

This isn't looking good, is it?

The neighbour rang on Friday to say that all was well and that, by the way, Kirsteen had had a calf. Lionel's immediate panic abated when it transpired that Kirsteen and calf had forded the stream. 'Up and about, then,' said a relieved Lionel. The next bit of news was less cheery: it looked as though Kirsteen had very little, if any, milk for the new arrival.

So on the Saturday morning of their departure I drew

them a map (not to scale, obviously) and sent them to Carrs in Penrith, secure in the knowledge that there they would be able to buy some substitute cow colostrum, rather than watering down replacement lamb stuff as they had done in the past. So few cattle remain in Hampshire and Berkshire these days that the local agri-sales outlets don't even bother stocking such stuff. It was a safe bet that they would come away with something else from Carrs that they had neither gone in for nor realised they needed until now, but even I could not have anticipated the impressive list of goods they accumulated. Lionel takes up the tale: 'We left you and found Carrs. What a fantastic place. Wish we had something like that around here! Having collected the colostrum substitute, a bag of calf milk powder, a special bag of dairy nuts for Kirsteen, who would probably turn her nose up at the sight of it, and a couple of Mag Rich Rockies, we left financially worse off than when we left you. This was going to be a very expensive calf and we still hadn't seen it – finally arrived home at 3.15 p.m. after a few short stops to stretch legs/change drivers/walk dogs, etc.

'Made up flask with substitute colostrum, found empty wine bottle' – like me, they only drink in the interests of recycling – 'and teat, put dairy nuts into feed bucket and set off in a westerly direction, wondering what we would find. Kirsteen was halfway up the hill and the rest were still on the other side of the fence. OK so far! As we approached, baby got up. Mum spotted the feed bucket and with tongue extending up each nostril in turn made a beeline for me. It met with her total approval, the best thing since last season's beet nuts! Peggy managed to get half the colostrum into the calf before Kirsteen left the now "empty" bucket and returned to see what we were doing. She can have the other half in the morning!

'The calf may be a little on the small side but appears quite lively and Kirsteen is always the very best mother, so they should be fine.'

There are definite parallels between Kirsteen's story and Mickey's in financial terms. They might become penpals, swapping tales of how best to separate their respective owners from wadges of folding stuff.

A week or so further on, another update arrived.

'Peggy's busy with the milk bottles once again. Kirsteen is getting very fussy about her nuts, especially when we ran out of the dairy nuts from Carrs. I'm afraid growers' nuts would not do, or the same ones with molasses poured over them, and certainly not rolled barley, or any combination of the three. No-one in this area has any stocks of dairy nuts, so today we went to one of the few remaining dairy farms around and asked the dairyman if he had any. No. "We don't use them, we mix our own," was the answer. Then it transpired that the farm had just received a load of sugar beet pellets. What? Kirsteen's favourite! And yes, we were welcome to a bag full. But would she still like them? Or want them?

'Tell me this: how did she know I was approaching with a bucket of sugar beet nuts? I was met at the gate. Hardly had time to get in. Her nose was in the bucket and no-one, but no-one was going to separate her from her tea.

'Have just heard that our local supplier is expecting a delivery on Thursday, so all is well with Kirsteen's world once again.

'Re next year's holiday. How about a week in July?'

And here's the best bit: they have named the calf after me. She is called Jackie of Ambrosine. I am delighted and honoured. There is a clear resemblance – we both have rusty-coloured locks – but her horns have yet to reach the same dimensions as my own. My husband says they extend quite alarmingly when the following things are mentioned: income tax, mobile phones, manicured countryside verges, designer label clothes for children, slugs . . . And her legs are far hairier.

There's a special affinity that binds creatures of the auburn sort: Highland cattle like young Jackie of Ambrosine, red squirrels and deer, Tamworth pigs, Irish setters, Janet Street-Porter, Nicole Kidman (though the presence of the last two on this list is somewhat intermittent) and me.

If you are out riding in the woods, it is perfectly possible to commune with the red squirrels and the deer as long as you do it in absolute silence, as once you are aboard a horse – ideally a chestnut one – the creatures of the forest become unaware of your existence, concluding that you and the steed are one. Wishful thinking, in my case, but it affords riders an intimacy that walkers and cyclists can never experience.

The Tamworth pigs I've already told you about, as well as their distant relatives the almost-but-not-quite wild pigs of Charlie Parker's up at Gilsland, but what's this about the Irish setters, I hear you ask. Will G&T not be mortally wounded by such a betrayal? I shouldn't think so, since G&T's loyalty is constrained by a certain pragmatism. Loyalty yes, but to whoever has the biggest biscuit/chocolate/steak trimming at the time rather than the touching, unswerving Greyfriars Bobby sort.

I have always secretly yearned for an Irish setter. I used to think it was because they were so engagingly daft but then a fellow I knew scrutinised me patiently and opined, 'Well, you would, wouldn't you?' Since the speaker was well over six foot four and had a Great Dane, he was hardly in a position to be cynical, but the euro definitely dropped: the issue of genetic preference was raising its red head again.

Of course, anyone who owns a dog is pretty much OK. Anyone who owns an Irish setter is better than OK. Ditto owners of collies, greyhounds, lurchers and absolutely any rescue dog. This irrational preference for doggy people is matched by an equal but opposite irrational prejudice against people who wear two-tone shoes and/or fixed grins; a slightly strange trio of maxims, you may think, but they have proved sound principles on which to base a contented, if uneventful, existence. On just one occasion, they clashed with cataclysmic effect.

Several years ago, my husband was away and working in particularly unpredictable conditions. Events escalated with terrifying rapidity and dramatic outcome; time, normality, life were in suspension. In the uneasy silence the telephone rang insistently. I picked it up.

'Don't worry, everything's fine. Everything is sorted.

All over. I'll give Malcolm a lift home – it's on my way.'

I didn't like the owner of the voice. I had dismissed him as a shallow, spivvy smart-ass primarily because he wore two-tone shoes; he didn't smile inanely, but those shoes were bad enough on their own: cream and tan, pointy toes and leather laces like rat tails. Fearful things.

'Right, thanks.' I could hear myself sounding disgustingly ungrateful – he had been working stupid hours too – so I ventured, 'Would you like to stay for something to eat?'

'No. That's kind of you, but my red setter has been missing me all week and I'd really like to get back to her.'

That shut me up. A man with a red setter that he needed to get home to – I could hardly hold anything against him, rat-tailed two-tone shoes or not, could I?

Red-haired women in history seem to possess a definitive spirit and elan. Elizabeth I was a pretty frisky filly and much further back, in something like AD 60, in the prehistory before even I was born, there was Boudicca. Acquisitive Romans were running about causing no end of trouble. In the east of England, Boudicca's husband King Prasutagus pegged out and left his own little empire between his two daughters and the Roman Emperor. An amicable settlement seemed out of the question, especially when a bit of rape and pillage ensued, defiling the daughters. Boudicca's feathers – red ones if legend is to be believed – were ruffled. First she raised an army and then she razed what she could of Colchester with a little help from her friends in a neighbouring tribe. Boudicca, like most redheads, was a bit of a bugger when in a bate. Torching Colchester was just a start: she proceeded to massacre its inhabitants, rout the Ninth Legion, tear down the temple, and go on a bit of a spree with a flamethrower; and once she was warmed up nicely, she sacked London for good measure.

Good old Boudicca.

Moral of this story: do not mess with redheads.

I have exported two sheep to Scotland. They have gone to Wick, after which you fall off. It has not been completely straightforward as they are the first sheep I have moved since FMD and a whole assortment of immensely complicated regulations have come in since then. And been changed. Then amended. Then cancelled. And so it goes on.

I heaped up all the directives from DEFRA about tags, forms and cleansing precautions. And then I tipped them all in the recycling box and phoned DEFRA instead.

The first call was aborted because I didn't press the right button and they didn't play Vivaldi while I waited. Everyone else does. Viv probably gets posthumous royalties up on his cloud for all the times his pieces are played between moronic announcements. I phoned again and the person I needed to speak to was 'out of the office at the moment'. The next time they were at lunch and the time after that I spoke to someone who said she might have an answer but couldn't guarantee it would definitely be the right one.

I could use my pink tags, with their individual numbers, she said. Almost definitely. I phoned back and yes, I could. Absolutely. So I filled out the forms and waited for William Armstrong from Longtown to ring to let me know the plan of action. And waited. And waited. And then phoned them.

'I left a message on an answerphone,' responded a slightly aggrieved voice.

Unfortunately, though, not my answerphone.

On Saturday, a cove called Derek from Armstrong's rang but I was in the bath and missed his call. I phoned back but he had 'gone out on the town'.

Next morning he was still there.

By the time I caught up with Derek on his mobile he sounded distinctly fragile so I made a few sympathetic noises and Derek recovered his composure sufficiently to tell me to deliver the sheep to Longtown by late afternoon the next day for onward transportation to Scotland overnight.

Freddie and I loaded them without incident and because I subscribe to that school of thought that says 'Trust in Allah but row away from the rocks' I took with me duplicates of DEFRA's forms, a year's supply of tags, and the gun (the *tagging* gun, just in case they caught an ear on some hitherto unnoticed protuberance during the journey and wrenched the existing ones out), and we delivered them to Longtown. I took all the paperwork to the guy in the Portakabin and held my breath.

'Aye, that's grand, lass. Nay problem.'

I very nearly kissed him.

I've got the new tags now, the yellow ones with the

sequential numbers, but personally I prefer pink for Manxes.

Winter has arrived suddenly. Mickey's presence means I need to be outside early each morning, feeding and mucking out. The sheep need shepherding and as I have split them into neat groups – feeders, ewes, tups – feeding too. The fires in the house have to be laid, fed, and emptied.

Lately it is not mellow or fruitful; it is just unseasonally cold and wet. Pebbles of rain ricochet noisily on the greenhouse roof at night; drizzle patters gently into the bird bath and early fallen leaves from the cherry tree are already pulping to softness on the lawn. Misty mornings precede cool days. Gold, ochre, umber colours glow under indecisive skies of steel grey and sharp blue. Torrents of rain leave sallow, sullen sage-green grass in their wake.

I suppose Keats was around long before global warming. That'll be it.

Our cottage season is slowing down now though it doesn't stop completely; many stoic souls regard a wintry wind and a sharp frost as ideal walking weather and a roaring fire at night as balm for the soul. I never regret our little bit of diversification although occasionally I am tempted to follow in the tradition of my granny's young and bright (female) solicitor who propped up a big sign on her desk saying 'No Soliciting' and call it cottaging. But, naturally, I daren't.

Other people have developed much more imaginative rural businesses but for us it had to be a cottage, since we didn't fancy farming snails – and apropos of nothing at all, I wonder who was the first person to look at a snail and think 'mmm, yummy' – couldn't offer education (about anything much at all, really) and were hopeless at painting other than the Dulux sort. To my considerable relief it seems that the guests are getting more eccentric each season. There may be hope for Planet Earth yet.

Apart from several springers, a lot of black Labs, some Jack-the-lad Russells, a greyhound who sat with her elegant paws crossed for most of the week and responded to the magic word 'walkies' with a look that clearly said 'Why, exactly?', the Runtweiler and the Dubious Dobermann, we had a charming couple with a police dog (failed). A gorgeous, loping hound, he was not aggressive enough to make the grade. Taking a liberal view of offenders, he frequently concluded that since they had done nothing to him he had no quarrel with them whatsoever and no, he would not be cajoled into bringing them to the ground, growling rabidly at them or showing them his dental appendages. He couldn't see the point. This was very much his private sojourn with his humans as the rest of the menagerie – twelve rats, an iguana, six guinea pigs, two other dogs – did not join them on this occasion. It gave them all chance to mourn the loss of the much loved boa constrictor.

It is a real privilege to meet people like that.

Visiting dogs like the cottage garden, which is walled on three sides, fenced on the fourth and therefore completely escape-proof. Except for a Staffie called Red who regarded the term 'escape-proof' as some sort of challenge and soared effortlessly over my left shoulder while I was chatting to her owners over the wall. Of all the problems I'd anticipated the hospitality industry might throw at me, I had rather overlooked that of low-flying dogs.

And then of course there is always the matter of poo. The town-dwelling populace just doesn't connect with this subject in the way we peasants do.

Sara and the gang were home for the weekend. Phoebe and Ellie were colouring, Sara's Malcolm was doing something he rarely gets chance to do (read the papers in peace) and Sara and I headed down to see to the cottage changeover. Sara noticed that the loo wasn't flushing efficiently so we got the plunger out and plunged . . . it was only marginally better and it was still gurgling menacingly so I thought I had better investigate the sewage holding tank.

Malcolm (mine) muttered something about being dramatic. Two women. One bloke. You know who was right, don't you?

The tank was full. Overflow imminent. We all peered into the slurry; I think we hoped for divine inter-vention but instead there was just a despairing silence, eventually broken by Sara, who observed, 'There's always sweetcorn in it, isn't there?'

The good news is that we captured one electrician and one builder, and together we pondered the alterna-tives . . . this is why, when there's a hole in the road, there are at least three people gazing into it. They too are thinking about alternatives. Our assembled little company cheered up no end when a quick flick through the Machine Mart catalogue identified a pump identical to the one we had just hauled from the bowels – and I use the word advisedly – of the pit. This was good. The fact that our pump also had something that looked suspiciously like a wig stuck fast into the bit where it sucks in the poo was not good at all. Our wonderful electrician extricated the obstruction and pronounced the pump 'knackered'. I think that is the correct technical term.

We prayed for a hold-up on the M6 to delay incoming guests and dashed into Carlisle and back clutching a shiny new pump. The electrician and the builder materialised as if by magic and in minutes it was fixed, and sucking away thirstily. The instructions included several sub-clauses about solids being no more than 35 mm in diameter and I will leave to your imagination how the combined imaginations of two sensible, educated, grown-up women reacted to this. We also devised a poetic entreaty, suitably framed and displayed on the loo cistern. It reads:

Please flush only paper and poo
Down this 'ere loo
We're not on the mains and our poor little drains
Cannot cope with other stuff
PS Especially no wigs

So far, this is working, even if we did run out of rhyming couplets at the end.

Mayhem receded just in time for the new guests' arrival. I painted on my best fixed smile and greeted the gentleman on the doorstep with a confident 'George Fitzsimmons, I presume'. As the guest's name was in fact Marc Allan, he went a bit pale and said, 'Oh, God, you're not double booked, are you?' Later that evening, they went to the pub – I'd booked them a table – where they were greeted with 'Good evening, Mr Wilson'. By then they probably thought they had landed in Royston Vasey. Still, no lasting damage was done and they joined us for supper later in the week. You will be relieved to hear that I managed not to poison them.

It is time to get some sunshine in the tank before our famed Cumbrian weather turns clashy. A few days away floating about on a boat will do the trick; the trouble is that this was all booked and paid for before Mickey came home and it feels altogether too harsh to pack him off to boarding school again. If I do that, he may decide he doesn't love me as much as I love him – even if admitting that you love a horse is to invite ridicule. Look at how history judged Caligula, a despot and a lunatic who spent many rewarding hours in conversation with the moon, yet is best remembered for having loved his horse so much that he made him a senator. As it is unlikely that history will bother with me in the same way as it does with mad Roman Emperors, it is probably safe to confess: I do love Mickey. His chiselled ears rarely waver from the upright position, he is happy, inquisitive and, unusually for a horse, positively affectionate, all just as his mother Kareima was. It would be impossible to do that horse whispering stunt of join-up with Mick because you can never get rid of him long enough for him to come back; there's something curiously complimentary about his closeness.

Mary, my riding companion, who probably did not know what she was letting herself in for, offered herself as sacrificial nursemaid, housekeeper and dung-bunger in chief. 'No problem,' she said, 'I'll look after him.' I warned her about Mickey's wholly unreasonable demands on room service, his pickiness about hay nets, adding that he rather liked a few succulents – carrots or windfall pears go down nicely – in his evening feed. I mumbled something about his inexplicable affectations about straight rugs and having his feet picked out in an unvarying sequential order. None of this put Mary off.

They got along famously.

To make things worse, my first trick on homecoming was to serve him dinner at breakfast time. Mickey summoned up a very withering look for a horse. It trumpeted: 'You can't get the staff these days.'

And the holiday was wonderful, even if it did have a catastrophic effect on my reliability with the feed scoop. Cruising is escape on a grand scale; there is a beautiful simplicity about sea and sky, something exquisite and perfect about miles and miles of absolutely nothing at all. Just frilled waves playing on a deceptively solid surface, concealing the tumultuous, many-tentacled, multi-finned world sluicing below.

Best of all, you only unpack once and they move the scenery for you.

On the first day of our holiday, I was mucking Mickey out at six in the morning and by six in the evening I was standing on board ship in a pair of silly heels looking out across Venice, that most magical of all cities. If I had a week's notice of being called to my Maker, my first words would be, 'Where's my ticket to Venice, then?'

In Taormina, on Sicily, we saw three ugly blokes in black suits, with black shirts and black ties, the first element of their ensembles rendering the second and third superfluous. If they weren't extras from a Scorsese film they were Mafiosi. I just didn't expect Mafiosi to go into ice cream parlours and order tiramisu cornets with a scattering of nuts. Perhaps they order extra threats to go, too.

This is a quiet ship. None of those annoying '*ding dong* World class tiddlywinks on the poop deck' calls. Any announcement that is made tends to indicate something either outstanding or dangerous. One in the first category informed us that we were to be passing the volcanic island of Stromboli at about 11 p.m. Oooer. So at 10.59 and a few seconds we went up on deck and watched Stromboli's silhouette draw closer and ever more menacing. Quite suddenly, small sparks fired upwards, glowing and angry, into the night sky. Then another plume of backlit blood red. More, taller spires of flame exploded like silent fireworks. And we felt the intense, furious heat. And smelt the molten lava, like a just-dead coal fire. A crater on Stromboli is very much alive.

A sensation, a wonder, an experience I shall never forget.

There was another, madder episode that wasn't in the brochure either. Travelling towards Tarquinia to see the Etruscan tombs, we passed a lumpen bit of roadkill. Nothing remarkable in that, I hear you say. Ah, but this bit of roadkill was a wolf. A big, rangy, shaggy wolf. You don't see many of those on the A6.

Every day I took a swim on the top deck: vital for figure and fitness and a time to think beautiful thoughts. Here's a beautiful thought: if swimming really is good for staying in shape, what happened to whales, sea lions and walruses?

Once I start those sorts of random musings, it is clearly time to go home.

Now I have put away foolish things, three inch stilettos and my little black dress (Tynedale retail park, £49 reduced from £279 – now that's what I call a bargain), until next year. I shan't be needing them shepherding through the Cumbrian winter.

As always, homecoming is a relief. The house hasn't burnt down, nothing is dead and there are no jackdaws trapped in the chimney.

Mickey whinnies cheerily next morning; an icy draught weaves under the door of the barn snatching at hidden crevices of warmth in my boots as I fill hay nets. Up on the hill behind the house, I can see across to Saddleback. High clouds, fresh out of dawn's starting gates, race one another across the wintry blue. When I am walking the dogs later in the day and looking across to the Pennines, the wind has lost its will and a single cumulus is impaled on a fell top. Today's landscape and last week's seascape vie for dramatic effect.

But this is home.

November – Full Circle

Sometimes, I look at Katie the lurcher and wonder what her past was like. Even if she could tell us, could I bear to listen? Then I ask her, 'Katie, are you pleased you came to live at Rowfoot?' She pricks her lynxy ears, sits upright stretching her front legs as far as they will go so that she looks like a meerkat and cocks her head on one side, and I know that she's saying yes. Lurchers – every home should have one.

THE YEAR HAD not finished with me yet. The Americans call it closure; I just call it the Bitter End.

There was none of that fancy feeling up and down the sheepy spine to see whether they were 'finished': in the old days we called it fat or grading but fat's politically incorrect now and graders were pensioned off and abolished years ago. When one's income depended on graders' whims, I used to give ours a flirtatious, girly smile in a pathetic effort to influence his judgement. And then he would glare at me.

No, this time it was much more straightforward. It was simply a matter of recalling what date the sheep were wormed, whether they were beyond the withdrawal period and when Freddie's pick-up was due its MOT. Once we'd got that lot synchronised, we could book them into the local abattoir, the three of them: Curly Horns who is nothing but trouble and leads the rest of the flock astray, The Triad, and the two-horned tup.

The Triad, with his uneven headgear of three horns, is a model possibly unique to Rowfoot who had four horns once but left one in a fence or a dyke or a victim somewhere. He had already attempted homicide or sheepicide or pesticide on his mate and then ploughed through a fence. As I dragged him back to his proper side it occurred to me that I might have stumbled upon the etymological genesis of the phrase 'battering ram'. Then again, perhaps not. Anyway, I fixed the fence with staples and a big hammer.

Two days later he did it again. This time he was in no mood to be captured. Blossom saw the bucket and, remembering how it felt to be starving, demonstrated her excitement by sticking her hoof through the hole he had made in the pig netting. She planted herself in the mud and gazed dreamily into the middle distance while I wrestled with her hoof – her newly shod, sharp hoof. Extracting it was not easy as Blossom was static and leaning on me. Blossom, you may remember, is no lightweight. With a mighty effort I thrust her leg back through the hole, but not before my hand had been trapped 'twixt hoof and fence. I swore. Quite a bit, I think. Then I chased the tup, fruitlessly, swore a bit more, gave up and walked down the field.

It was then that I noticed my hand. Even in the dusky half-light I could see the upper surface engorging quite terrifyingly with blood. A trickle ran down my forefinger and the pain was truly intense.

Blood bothers me. Years ago, at Hickstead of all places, I mistook a finger for a piece of melon – easily done, I'm sure you'll agree – and sliced through it. I passed out cold. When I came round, some kind soul was asking me whether I needed anything, and with a commendable sense of theatre I whispered, 'Brandy, brandy.' A rather good glass of Remy Martin appeared. I have favoured this course of action ever since, but on this occasion there was little time to waste quaffing premium brand spirits as the doctor's surgery was about to close.

I came home with a dressing on my left hand, a sling to drain the blood back down my arm, and a sore right arm from a tetanus jab. At least the pain was nicely balanced.

It is not easy sleeping with your arm suspended from the ceiling but manageable in short bursts.

Next morning Two-Horn Tup had his head stuck in the gate. Sheep hang themselves like this. But it is a reasonably new gate and a reasonably mature sheep, so I fetched the saw and the shotgun and sat on the wall and thought about it for a bit. Then I went and dragged Malcolm out of bed. He is more spatially aware than I and he managed to corkscrew the tup's head round and free him. Then he swore imaginatively at the sheep and went back to bed. The tup was last seen covering his ears with his cloven hooves and pleading for clemency.

It was turning into a bad weekend.

Next, we found a puncture on the car, probably inflicted during yesterday's mercy dash to the doc's. By now patience and new swear words were things we were running out of.

By the end of the day I had discovered an outcrop of what looked very like orf on my face, my arm was swollen from the tetanus, my hand hurt from the bruising and the prospect of life in a city centre terraced house with only a goldfish for company seemed a highly attractive alternative to living amongst a horde of demented, suicidal animals. We would need a corner shop so that the papers were handy at times of punctures too, of course.

I moaned. Malcolm said he was thinking about a trade-in. I think he meant the car but he might have meant me as I am becoming tiresomely grumpy in my old age. He has talked about wife swapping in the past but he has yet to find someone willing to take me off his hands. Also, his idea of a good swap would be me in exchange for a cordless hammer drill, which leads me to conclude that he has not really grasped the principles of wife swapping at all.

Anyway, come Tuesday, Curly Horns and two tups are taking a one way trip out of here, just before the MOT runs out on Freddie's pick-up. Brilliant timing. The only down side is that I have been told that I can't have the horns back because they are something called SBM. SBM? I thought I had their Greatest Hits CD. But, drearily, SBM turns out to be Specified Bone Matter and as someone in Brussels worries that it might cause me terminal damage, I am not allowed to have it. It might, if I were going to roast, stew or fry them, but horns being such unpromising gastro-fuel this was not my intention. I had always hoped that The Triad's horns might have made a crook handle but this, like so many dreams, will be for ever unfulfilled.

The meat will be cut up for the freezer and I hope the freezer is suitably impressed. It is a good thing that the Rayburn is fired up as we may need to braise some of it for several weeks even to make it into dog food, but at least the evening shepherding round will be a stress free zone.

And all this chimes nicely with Prince Charles's initiative to get us eating more mutton. His Royal Highness has been silent on sweetbreads thus far, but if he saw those on the tup, I am sure he would find something to say on the subject.

All we had to do now was clip the sheeps' tummies. This seems a tad pointless when they are off to be killed – I mean, you wouldn't go to the hairdresser's on Monday if you had the gallows pencilled in for Tuesday, would you? But apparently it's to do with E. coli bugs sneaking from hairy belly into the chop department and it's another of the Things You Have To Do. So we did it: sat them on their backsides and set the clippers a-whirring. One tup was having flashbacks to clipping and nearly dewillying time, I fear, and he sat very still indeed, only going into cross-eyed and traumatised mode when Freddie got round to denuding his balls.

Freddie and I took the sheep off without incident and nearly a fortnight later – mutton needs serious hanging – we collected the joints and chops from the abattoir. And oh, they do look good. Just so as not to tempt providence, I stowed some in the freezer in one of those blue bags the district council give us for rubbish. Blue for a boy – and it could still turn out to be rubbish, of course. The dogs are hoping so. A hind leg of the old ewe has been boned out and cured in readiness for smoking. It could be completely inedible and useless except for resoling shoes. But it's worth trying.

My stars this week said it would be a week full of incident, that aspects of my life would be substantially enhanced – maybe the astrologer was thinking about the dimensions of my left hand when he wrote this – and that my lucky biscuit was lemon puffs. I'm off for a fig roll now. I don't trust myself with lemon puffs.

It's taken us thirteen years but we have finally found a useful role for Tess. She has an unerring ability to sniff out rats. Thankfully, it's a little used talent. Like everyone else, I've heard – and cringed at – that statistic that you are never more than six feet away from a rat. This is fine if the six feet is measured when I am out in the backstreets of Carlisle but it is not fine at all when the six feet is measured somewhere in my own back yard. This morning Tess has been paying very close attention to a crack in the concrete at the side of the gig-house door, the white tip to her tail quivering in expectation, her eyeballs popping and her ears sticking up like a pair of Jodrell Bank misshapes: a most peculiar sight.

I relayed all this to Malcolm who just said that he thought Tess was going senile.

I said that he might very well be right but I still thought we had better look inside. But I can never cope with the intricacies of the lock on this door: it is one of those designed to stop burglars getting in but it lacks the discriminatory capability of admitting the rightful owners, so Malcolm had to come and fiddle and wiggle it. Then the key broke and we spent a fun five minutes trying to extricate the stump from the hole, a procedure that involved a pair of pincers and a lot of WD40. After a search for the spare key and a second attempt to unlock the door we finally reached the stage of mission accomplished, finding three half-grown rats asleep under the pony trap. No parents in sight: that's a sign of the times.

Now, three days later, we have a body count of three – almost as many as an episode of *Taggart*. One natural causes, one axed to death and the third poisoned. The axing was a ticklish job: on the second morning of Operation Rat, Malcolm instructed me to fetch the axe. 'No, I tell you what,' I said, 'you take the axe and I'll do the door.' It was too early in the day for combat, to my mind. And you know what? The crafty little rat disappeared – or thought he had disappeared – under the sheet that covers the seat on the exercise cart. However, he left his tail hanging out, betraying his position precisely. I need say no more, I am sure.

The postie arrived mid-rodenticide. He was appalled. 'Orphan murdering. You should be ashamed of yourselves.' Then he smiled broadly and went on his way.

*   *   *

Anyone who is owned by a dog knows that dogs insinuate themselves into every secret corner of your life, that they never judge you, that they – even, or perhaps especially, when all the world is against you – remain on and at your side. Their lifespan, though, is shorter than our own so losing a dog is an appalling inevitability.

So it was with Gyp. A series of terrifying fits left her unrecognisable as the dog we had known and loved for thirteen years and us knowing what we had to do. That the debt of gratitude we owed her was best settled by releasing her from suffering was beyond doubt and for the thousandth time we gave thanks for our wonderful vets and nurses who cared for her and, indeed, for us. And yet, and yet . . . we still felt like murderers. We buried her in the garden and planted gypsophila there – an obvious choice, gyp for a Gyp – and daffodils, which next spring will sway as gently as her nature, and we watched, late one night, as a barn owl flew silently over her garden-grave like a blessing, its wings kissed with moonlight. We miss her. She was a rare dog, uncommonly tolerant, innately loyal and profoundly affectionate.

Phoebe, my younger goddaughter, came up with a touchingly graphic tribute to her lost friend: a representation of heaven as Phoebe saw it – God, all beard and age, the Pope, and Gyp. It's good to know that the Pope is in such good company.

Normally, I like a decent interval to elapse between the loss of a dog and the acquisition of another. In time, you come to terms with needing and finding another dog; what you cannot do is 'replace' a dog – they are not broken plates. But this time things were complicated by Tess, who seemed bereft and listless in solitude. G&T having been litter sisters, Tess had never been an only dog. Her desolation made finding a new special friend for her a matter of some urgency. Not a puppy, we decided. It would torment her. It would have to be something possessing her own Bony Maronie physique: no smaller than a whippet, no larger than a greyhound. Since I have long abandoned hope of finding a collie that can work these Manxes, its breed hardly mattered.

Slightly uneasily, we set off for the Animals' Refuge at Wetheral. We were interviewed to see if we met the criteria for responsible parents before we were allowed anywhere near the candidates (a scheme that has much to commend it for humans but is, I fear, unlikely to catch on) and Tess came with us so that she could interview the applicants. Well, it was either that or an advertisement in *The Bitch*: *Elderly collie needs companion. Waggy tail, agreeable disposition essential. Full domestic staff kept. Own bed but share of dog house (between the Rayburn and the gas stove) considered in longer term. Remuneration by arrangement; generous Bonio ration.*

There was, we were told, a 'lovely little lurcher' in the kennels. Our interviewer may have mentioned an extra incentive of a free flat cap and two ferrets or I may have made that up: sure enough, in kennel two there was a little brindle and white lurcher bitch that looked like a body-double for Alfred, the canine star of *Heartbeat*. Rakish, with a sharp Mohican haircut, strawberry pink paws and a straggly beard where she could save bits of dinner for Ron, LaterRon, bits of her were so thin that you could see right through them. She had no name, just a number.

Tess wagged her tail joyously at meeting her and together they pulled on the two-pronged lead in perfect time. I'd wager that the first person to exclaim 'Look at the arse on that' was behind a lurcher on a two-pronged lead when he did so, possibly on a skateboard being hauled along like some latter-day Ben Hur, as this otherwise dainty lurcher had the most muscular backside I'd seen off the rugby field.

She has, you will have guessed by now, come to live at Rowfoot and charmed us utterly. Next, we had to name her. Dog 151 would not do. I suggested Maureen but that was vetoed. Jess would have suited her nicely but that way lay an identity crisis: Tess would come when Jess was called and vice versa. More likely, no dog would come at all, Tess being pretty deaf these days and New Dog having no idea that she was called Jess anyway. She ought to have a pretty name, a noble name, I said. Lucinda, perhaps? Nicely aristocratic, and it could be abbreviated to Lucy. Nothing stupid or insulting like Malcolm's unkind suggestion: Scruff.

We settled finally on Katie, partly in homage to my goddaughters' affection for the Katie Morag stories and partly to acknowledge the part played in all this by Katie the kind veterinary nurse who suggested we might usefully make a trip to Wetheral. Katie Morag she is and it suits her very well.

Katie can complete a circuit of the back field in

minus two seconds and do 0–60 in even less. Tess watches, with a look that says, 'I'll hold your coat.' Katie Morag has her own chair too: our initial outrage – we have never permitted dogs on furniture before – softened when we considered a) in the great scheme of things, how important is an old chair? and b) when you are that bony, even sharp in places (not a condition I can identify with, upholstered as I am), how can the floor be a comfortable place for proper repose? A lurcher *needs* a chair, obviously. She can eat upside down which is quite clever for a dog and sleeps anywhere at all, but her favourite place is on Malcolm's lap, her bottom towards the rest of the company in the sitting room and her head tucked in between his lower arm and that of the chair.

If either of us goes upstairs to inspect the porcelain, she dances a mad jig when we come down again; we have explained that we love the welcome, that we are grateful that already she loves us, but that for brief trips away the short welcome – ears pricked and a casual wag of the rudder – will suffice. She takes no notice. I've bought her one of those extending leads and told her not to garrotte Tess; instead she has found it most amusing to truss me to a tree and grin while I unravel the knitting. Her masterstroke is the Sock Trick. As we leave the house, she pulls a sock out of one of my wellies by the back door and stands with it dangling from her mouth as if to say, 'You can go, but you can't go far, not while I've got one of your socks.' It is probably wiser to keep quiet about the drawer full of footgear upstairs lest we worry her.

We will never forget our gentle Gyp. She has not been replaced. But we have another dog – another unique, rewarding and very special dog: Katie Morag. Long may she stay.

Epilogue

N OT MUCH HAS changed throughout this last year. My sheep, like most in Britain, still run about on an extensive grazing system. Sheep, you see, do not respond well to intensive systems that restrict both their diet and their freedom to sneak through fences and scale walls at will. As with other sectors of the livestock industry, the organic and rare breed market is burgeoning, although delicacies such as Manx Loghtan lamb and Herdwick mutton need careful husbandry in life and careful butchery afterwards.

I wear my own sheep: my woolly Manx hat is completely waterproof and protects my head from damp rising from soggy wellies; but most of the annual national clip is sold at a depressingly risible price and, due to ridiculous rules, the Wool Board both controls prices and fails to develop unusual products that might prove a lifeline for shepherds in the uplands. Captains of almost any other industry would invite the Monopolies Commission round for a quiet chat and a glass of the hard stuff, then send them off to take up the cudgels with the Wool Board. The Wool Board has suffered no such inconvenience and has not only escaped public flogging, but still persists with its restrictive practices.

A few crafty artisans manage to slip through this particular net and in the Cumbrian uplands a group of ladies calling themselves the Wool Clip run a co-operative and a shop selling threads, hats, jumpers, slippers, belly button warmers and heaven knows what else whilst staying within the bounds of legality. Even if they were to slink across a few pettifogging legal thresholds, they would probably still be OK because they look harmless, wear tweeds (well, what else would you expect them to wear? Polyester? I think not) and speak softly.

Perhaps we in Cumbria are so remote from the urban hubs that our transgressions are perceived as too minor to have an impact. Perhaps city sophisticates even think we are a teensy bit irrelevant.

Elsewhere in Cumbria, an enterprising young lady called Christine Armstrong took the old idea of lining farm lofts with surplus fleeces for insulation, rethought it, updated it and, with a strange alchemy of pixie dust and clever new twenty-first-century technology, created Thermafleece. Made from the woolly jumpers worn by the rather well dressed Herdwick sheep just outside her back door, Christine's supremely innovative product keeps National Trust houses up and down the land warm in winter and cool in summer.

So, sometimes we peasants are irrelevant. And some-times we are surprisingly innovative and smart. It's a puzzle. And it baffles our metropolitan peers no end but we quite like that because it keeps them on their toes.

I'll stick with my mad, edible, eminently wearable sheep for now. I wouldn't want to be a dairy farmer again because I was a Direct Seller and if I were one of those now I would never sleep. Instead, my nights would be filled with desperate dreams peopled by wraiths waving writs and claiming I had poisoned them with my cheese. It's not just an old wives' tale that cheese causes nightmares, you see. Not that the mainstream boys are any better off, shackled by quotas and seemingly powerless to arrest collapsing milk prices that have occurred since the death – some might say murder – of the co-operative known as Milk Marque. Beef farmers fare just as badly. The architects of the damage are, in no particular order, BSE, intensive farming practices, foot and mouth (unfair, but there it is), and imports of cheap – or at any rate, cheaper than home-produced – meat from countries where foot and mouth is endemic. Strange, but true. And that's just a start. As a list, it is by no means exhaustive.

I don't think I could be lured back into rearing pigs either as I really prefer not to have known my pork personally before eating it. I must be getting soft in my old age. Pigs are at least sticking to what they do best as few have troubled Equity lately, Babe having grown up and left home and others with theatrical ambitions confining themselves to the small screen – *Ready Steady Cook* mostly. Pigs don't like being intensively farmed any more than sheep; sometimes they respond to the hostility of their environment by eating one another. They are nothing if not smart, so no doubt one will soon work out that the way to assure their place in history rather than on daytime TV cooking programmes is to eat their keepers instead.

Nor do I envisage another goat, although I would never turn away a Goat King taking a walking holiday in the Cumbrian fells. The Irish, with their celebrated uncanny instinct for eccentricity, crown a wild mountain goat King of Ireland for three days; his rule over the Puck Fair street festival in County Kerry is a fitting tribute to the wild goat that alerted Killorglin town to Oliver Cromwell's uninvited presence in the seventeenth century. If an Irish Goat King honoured me with a royal visit I would naturally install the appropriate blue plaque. It would be ungrateful not to. But no permanent goats.

I have nothing useful, instructive or worthwhile to say on the matter of horticulture as plants just don't do it for me. I make an effort with a vegetable patch, really I do, chiefly because it is a terrible shame to let all that expensively cultivated manure go to waste. I like the idea of growing my own but I don't like the weeding it involves.

I grew a fantastic crop of potatoes once. A small boy accompanied me for the ritual picking for supper. He viewed the process of digging and shaking with considerable suspicion.

'What you gonna do with them?'

'They are for dinner,' I replied brightly.

'My mum gets 'ers from Sainsbury's.'

My bright mood dimmed slightly.

"Ers are cleaner than them 'uns.'

People talk to plants but I find them poor conversationalists. A dog will waggle its ears, affecting interest. A horse might swish a tail. A sheep will disappear in the opposite direction, but look on the bright side: at least it's a reaction. But a plant? Nothing. Not a flicker. Very unsatisfactory, but if you come across a decent recipe for chickweed, be sure to let me know.

So it's a case of horticulture no, but horsiculture yes. Still girlishly in thrall to the species equine I remain, especially since Mickey has come home. It is looking less and less likely that he will be setting off in search of pastures new as he quite likes the ones he has already. After protracted negotiations we have resolved his early retirement issues and pension package: he gets unlimited feed and sleeps on the latest thing in equine bedding (rubber mats), he has a pedicure and fresh footwear every two months and in return he carts the old girl around the parish a few times a week and looks gorgeous: he reckons it's a fair deal. I think I may have been conned.

The bigger picture for farming remains an impressionist fog. There is a bright future for hobbyists with funds to subsidise their stock keeping. There might even, at a pinch, be a future for proper farming. But yokels need not apply. Modern farmers need to be seriously savvy, have a working grasp of marketing nous and computing skills, retain the usual deftness with a hammer and a six-inch nail and have a bit of spare change in the back pocket. As Malcolm's uncle, a retired farmer himself, advised us when we first came to Cumbria, back in 1982: 'Bring your townie money, lad. There'll be plenty farmers ready to take it off you.'

Some wise soul once asked, rhetorically I think

though I cannot be sure: 'How do you make a small fortune in farming?' The same wise person replied, 'Start off with a large fortune,' giving a creepy little snigger whilst delivering the punch line. He might merely have been incurably cynical. He certainly took a bit of a risk by talking to himself – people have been borne away to a place of safety for less – but on the whole, it is not bad advice.

Some financial wizard worked out that a farmer's pay, taking into account his (or in my case her) total man-hours, worked out as, oooh, as much as 50p an hour. Times is 'ard, you see, farming, you see. But you can keep a few sheep without being a farmer, as long as you're content to be a smallholder, stranded in an agricultural netherworld 'twixt 'proper' farmers – they're the ones with proper tractors and improper overdrafts – and enthusiastic hobbyists who are something in the City with clout, clean hands and unlimited funds. Smallholders, like many of the stock they keep, are a rare breed. Here's how to spot them.

- They demonstrate an irrational level of commitment to their livestock, which are frequently ugly or weird or both.

- They have at least a couple of other jobs, and lavish the majority of their earnings from gainful employ-ment on their peculiar animals from which they make no money at all.

- They grow hides like rhinoceroses to withstand the scorn of 'proper' farmers and the jibes of their peers who still have sensible jobs like accountants and personal trainers.

- Unlike farmers, accountants and personal trainers, smallholders see the funny side of most things involving muck. They have to: it's the law.

And, reader, I am still one of that happy breed.

And now that it's November once again, it is time to put the tup in with the ewes. And that is just about where we came in, isn't it?

THE FUNNY FARM
by Jackie Moffat

We often talk about leaving the bustle of metropolitan life behind and going in search of pastures new but rarely do so. Jackie Moffat is one of those who did.

It was in 1982 that she and her family, armed with a bucketload of optimism, stout books and highly developed sense of the ridiculous, bid farewell to the London suburbs and headed north up the M6 to Cumbria. Their destination was Rowfoot, a small, dilapidated dairy and stockrearing farm (although mice seemed to be the only stock in evidence on their arrival) nestling in the idyllic and celebrated Eden Valley. Their intention was to start leading 'The Good Life' and get to grips with the reality of running a working farm. After over twenty years of learning the rural ropes – and especially the vagaries of the farm's four-footed residents: the sheep, cattle, pigs, horses, dogs, not forgetting Millie the goat – Jackie and Rowfoot are each going strong, concentrating on rearing Manx Loghtans, a rare breed of sheep originating from the Isle of Man.

Inspired by her column in *Cumbria and Lake District Life* magazine, *The Funny Farm* is Jackie Moffat's funny, wise, heart-warming and at times moving account of the day-to-day trials, tribulations and triumphs she's experienced – the story of a woman at one with her life even if on occasions, she feels completely at odds with the rest of the world!

9780553816556

BANTAM BOOKS

C'EST LA FOLIE
by Michael Wright

One day in late summer, Michael Wright said a fond
farewell to his comfortable South London existence and, with
just his long-suffering cat for company, set out to begin a
new life. His destination was 'La Folie', a dilapidated
15th century farmhouse in need of love and renovation
in the heart of rural France . . .

Inspired by the success of his much-loved Daily Telegraph
column about La Folie, this book is his winningly honest
account of his struggle to make the journey from chattering
townie to rugged, solitary paysan. In chronicling his
enthusiastic attempts at looking after livestock and coming to
terms with the concept of living Abroad Alone, the author
gradually discovers what it takes to be a man at the beginning
of the 21st century, especially if one is short-sighted,
flat-footed and not much good at games.

Life-affirming and laugh-out-loud funny, Michael Wright's
tale of a new-found life in France with a cat, a piano and an
aeroplane, is as much an elegy for rural France as a hymn to
the simple pleasure of being alive.

9780553817324

BANTAM BOOKS

ON A HOOF AND A PRAYER
Around Argentina at a Gallop
by Polly Evans

At the age of thirty-four, Polly Evans decides to fulful a childhood dream – to learn how to ride a horse. But rather than do so conveniently close to home, she goes to Argentina and saddles up among the gauchos. Overcoming battered limbs, a steed hell-bent on bolting, and an encounter with the teeth of one very savage dog, Polly canters through Andean vineyards and gallops beneath snow-capped Patagonian peaks. She also survives a hair-raising game of polo and a back-breaking day herding cattle.

Taking a break from riding, Polly delves into Argentina's tumultuous history: the European's first terrifying acquaintances with the native 'giants'; the sanguinary demise of the early missionaries; and the gruesome drama of Evita's wandering corpse.

On a Hoof and a Prayer is the stampeding story of Polly's journey from timorous equestian novice to wildly whooping cowgirl. It's a tale of ponies, painkillers and peregrinations – not just around present-day Argentina, but also into the country's glorious and turbulent past.

9780553816792

BANTAM BOOKS

NARROW DOG TO CARCASSONNE
by Terry Darlington

'WE COULD BORE OURSELVES TO DEATH, DRINK
OURSELVES TO DEATH, OR HAVE A BIT OF AN
ADVENTURE . . .'

When they retired Terry and Monica Darlington decided to
sail their canal narrowboat across the Channel and down to
the Mediterranean, together with their whippet Jim.
They took advice from experts, who said they would
die, together with their whippet Jim.

On the *Phyllis May* you dive through six-foot waves in the
Channel, are swept down the terrible Rhône, and fight for
your life in a storm among the flamingos of the Camargue.

You meet the French nobody meets – poets, captains,
historians, drunks, bargees, men with guns, scholars, madmen
– they all want to know the people on the painted
boat and their narrow dog.

You visit the France nobody knows – the backwaters of
Flanders, the canals beneath Paris, the heavenly Yonne, the
lost Burgundy Canal, the islands of the Saône, and the
forbidden ways to the Mediterranean.

Aliens, dicks, trolls, vandals, gongoozlers, killer fish and the
walking dead all stand between our three innocents and their
goal – many-towered Carcassonne.

'WRITTEN WITH THE AUTHOR'S GLORIOUS SENSE
OF HUMOUR, THIS IS ONE OF THOSE JOURNEYS YOU
NEVER WANT TO END'
The Good Book Guide

9780553816693

BANTAM BOOKS

TEACHER! TEACHER!
by Jack Sheffield

Miss Barrington-Huntley took off her steel-framed spectacles and polished them deliberately. 'Mr Sheffield,' she said, 'after careful consideration we have decided to offer you the very challenging post of headmaster of Ragley School'.

It's 1977 and Jack Sheffield arrives at a small village primary school in North Yorkshire. Little does he imagine what the first year will hold in store as he has to grapple with:

Ruby, the 20 stone caretaker with an acute spelling problem

Vera, the school secretary who worships Margaret Thatcher

Ping, the little Vietnamese refugee who becomes the school's best reader and poet

Deke Ramsbottom, a singing cowboy, father of Wayne, Shane and Clint,

and many others, including a groundsman who grows giant carrots, a barmaid parent who requests sex lessons, and a five-year-old boy whose language is colourful in the extreme. And then there's beautiful, bright Beth Henderson, a deputy head, who is irresistibly attractive to the young headmaster . . .

'*Heartbeat* for teachers'
Fay Yeomans, *BBC Radio*

9780052155281

BANTAM BOOKS